THE NATURAL
VEGETATION
OF NORTH AMERICA

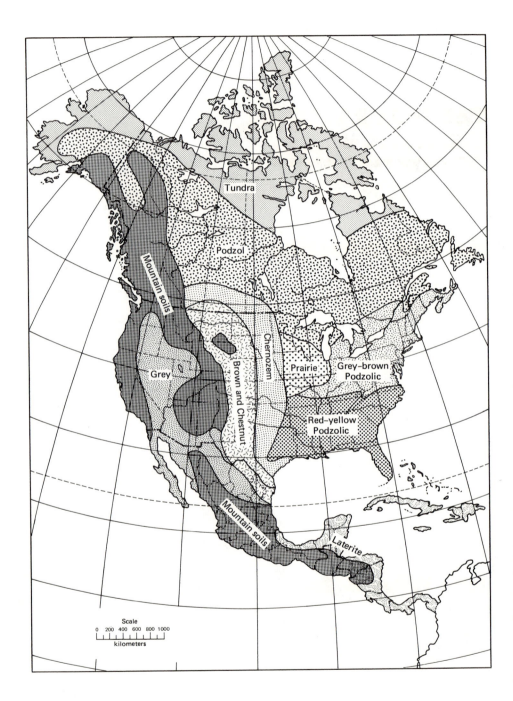

Tundra

Podzol

Mountain soils

Chernozem

Grey

Brown and Chestnut

Prairie

Grey-brown Podzolic

Red-yellow Podzolic

Mountain soils

Laterite

Scale

0 200 400 600 800 1000

kilometers

THE NATURAL VEGETATION OF NORTH AMERICA

An Introduction

John L. Vankat
Miami University

John Wiley & Sons

New York
Chichester
Brisbane
Toronto

To my family

Library of Congress Cataloging in Publication Data:

Vankat, John L   1943-
  The natural vegetation of North America.

  Includes bibliographies and index.
  1. Botany—North America—Ecology. 2.   Plant
communities—North America. I. Title.
QK110.V36   581.5'097   78-31264
ISBN 0-471-01770-1

Printed in the United States of America

10 9 8 7 6 5 4 3 2

Preface

The purpose of this book is to acquaint the reader with an important aspect of the natural world. With the widespread environmental interest and concern of the last decade, it may appear trite to say that a book is written to help readers better understand the environment. Numerous, varied texts with that purpose are already available. Some of these concentrate on the fundamentals of the science of ecology; others emphasize environmental problems. A complementary approach describes and explains biological systems, such as deciduous forests, deserts, grasslands, and so on. This book has as its theme the study of vegetation—the dominant component of these biological systems. In addition to providing a basic understanding of vegetation and its study, I hope that this book will also develop the reader's awareness and appreciation of a natural heritage that is continually being eroded as a result of human population growth and resource consumption.

The book is divided into two sections. Part 1 outlines the basics of vegetation science as they apply to North America. Its organization stresses a holistic approach, that is, one in which all aspects are interrelated. No single topic is covered in great detail, but a list of suggested readings for further study is provided at the end of Part 1. Part 2 constitutes the main focus of the book. It covers the major kinds of terrestrial vegetation of North America. Each individual type of vegetation is considered in detail and is interpreted in terms of the background material of Part 1. Extensive use of illustrations permits readers to visualize vegetation types that are unfamiliar to them. In Part 2, a suggested readings list is given at the end of each chapter.

This book is intended for two types of university courses. At the upper-class level it would be useful in ecology and biogeography courses and in most cases would be supplementary to another text. Some instructors may choose to use this book as the sole text by adding detailed readings and/or lectures to expand the material in Part 1. The other courses for which this book is intended are freshman and sophomore level courses that study the description and explanation of biomes. Depending on student backgrounds, the instructor may need to initiate the course with a few lectures on general plant biology. Instructors teaching at any level should find it relatively easy to incorporate material on zoological topics if a more inclusive biological approach is desired.

Finally, many people have aided me in a variety of ways in the preparation and completion of this book. To name a few would be to slight many. Nevertheless, I thank them all. The entire project has been exciting for me, and I sincerely hope that readers will be able to share in my interest and enthusiasm for this subject matter. Suggestions for improvement are most welcome.

John L. Vankat

Contents

PART 1

THE STUDY
OF VEGETATION
AS APPLIED
TO NORTH AMERICA

1

INTRODUCTION

Vegetation dominates much of the appearance of the world's landscapes and greatly influences human activities in many regions; yet, its importance is surprisingly unrecognized. This discrepancy is a product of the great physical and mental separation between people who live in technological societies and natural environments. Nevertheless, everyone has vivid mental pictures of vegetation. Most can envision the towering forests of conifers along the Pacific Coast, the uninterrupted expanses of grasslands in the Great Plains, the brilliant colors of the fall foliage in the East, the desolate winter appearance of the tundra in the arctic, the dramatic flowering of the cacti in the deserts, and the lush density of the rain forests in the tropics. Consequently, we all have an intuitive feeling for vegetation, and this can be used as a basis for the development of our knowledge about it.

Most people typically use the term vegetation in a very broad sense as a description of plants in general, be it the potted plants of a home or the Everglades of southern Florida. However, the term has a narrower meaning to plant ecologists; **vegetation** is defined as the plant cover of an area. Thus, there is the coniferous forest, the grassland, the deciduous forest, the tundra, the desert, the tropical rain forest, and so on.

Referring to vegetation as "natural" presents problems of definition. Some scientists refer to natural vegetation as that unaffected by humans. Others mean the vegetation that would form if the human species became extinct. These definitions have two shortcomings. First, there are no objects of study, since all vegetation has, to some degree, been affected by humans and obviously the species is not extinct. Second, they denote that all human activities are unnatural—a debatable, philosophical proposition.

In this book **natural vegetation** is defined as vegetation that would be in existence today if the permanent European settlement of North America had

3

FIGURE 1.1   Some of the many flowering plants in a Tall Grass prairie in late summer.

not taken place. Such vegetation approximates that of presettlement times (assuming that changes since 1607 would have been minor), so this definition provides objects of study, making it less subjective than the others. However, this definition of natural vegetation is not used because of some inherent "goodness" in presettlement conditions, nor is it used because the presettlement environment can be considered to have been "untouched." In fact, the aboriginal Indians had a significant effect on the vegetation of most areas but, in contrast to that of postsettlement times, their impact was relatively constant for centuries because of slow growth in population and technology.

Vegetation and its study are important in many ways. There are practical applications that must be considered in the production of wood, forage, and wildlife; thus, forestry, range management, and wildlife management all involve vegetation study. Forestry, in fact, is basically a specialized vegetation science. Vegetation study is also essential in the conservation of resources such as soil, water, and even genetic material. Vegetation is used to reduce soil erosion, manage watersheds, and store the gene pools of species populations. Furthermore, since vegetation is a product of the environment, it frequently can be used as an indicator of environmental factors. For example, since different kinds of vegetation are often associated with different soil types, the preparation of some soil maps is greatly aided by observations of vegetation, in place of the laborious, intense sampling of the soils (Figure 1.2).

There is also the important relationship between vegetation and recreational activities such as hiking, camping, skiing, bird watching, and flower observation. In addition, the aesthetic value of vegetation is incalculable; it has been the inspiration for great works of art and can have a calming, settling influence on the human spirit.

The importance of vegetation and its study is also exhibited in the influence that vegetation has had on the history of the human settlement of North America—in the aboriginal Indians' use of woodlands, prairies, and other areas, in the clearing of eastern forested land, in the plowing of midwestern grasslands, in the lumbering of western forests, and in the developing awareness of our environmental heritage as expressed in the establishment of national, state, and local parks and preserves.

Before discussing vegetation science in more detail, it is necessary to

FIGURE 1.2   An example of where vegetation can be used as an indicator of soils. The coniferous forest in the background and the shrub desert in the foreground are restricted to different soil types.

present a few important terms: first those terms for organisms in general and then those for vegetation in particular. Ecology is a word that with recent popularity suffers from misuse. A traditional and correct definition specifies **ecology** as the study of the interrelationships of organisms and their environment. Aspects of the study of populations come within the scope of ecology. A **population** is a group of interbreeding individuals of the same species; examples include the sequoia trees of the Giant Forest sequoia grove in California, the insectivorous sundew plants of Cedar Bog in Ohio, the dandelions of your neighborhood, and the wolves of Isle Royale National Park. Populations, in turn, make up communities; a **community** is the group of individual organisms inhabiting an area. The term plant community refers only to the plants present; the term biotic community includes all organisms. Giant Forest has many other plant species besides the sequoia trees; collectively they are a plant community. The same is true for Cedar Bog and neighborhood lawns. The wolves of Isle Royale form a portion of a biotic community, which includes all organisms found there.

A still more inclusive term is that of ecosystem. Systems in general are assemblages of interrelated, interdependent components; in ecosystems the components are either biotic or abiotic. The biotic components of an ecosystem make up the biotic community, that is, include all the organisms. The abiotic components are those that are not alive; they are classified as matter or energy, and examples include the atmosphere, solar radiation, and soil minerals. **Ecosystems,** therefore, are collections of organisms and environmental factors as they are interrelated by the exchange of matter and energy. Isle Royale, being surrounded by Lake Superior, is a relatively well-defined ecosystem, but Giant Forest, Cedar Bog, and lawns can also be considered ecosystems, albeit ones with less distinct boundaries.

Just as the terms population, community, and ecosystem form a series of increasing breadth, so do the following vegetation terms: stand, association, and formation. A **stand** is any example of vegetation. Giant Forest and Cedar Bog are stands, as is a lawn. An **association** is a group of stands with similar characteristics. For example, there is the giant sequoia grove association. It includes all forest stands containing giant sequoias, since those stands are much alike. The term association is defined and used in various ways; be careful of the meaning when reading other books and articles on vegetation.

Different but somewhat similar associations may be grouped together in formations. A **formation** is a group of associations that are dominated by a similar form of growth. For example, the giant sequoia grove association is characterized by coniferous trees. So is the white spruce-balsam fir association of Canada and the ponderosa pine association of Arizona and elsewhere. These and other such associations collectively make up the coniferous forest formation. The chapter headings in Part 2 of this book list major formations of North America. Formations may be referred to as biomes, if animals and other organisms are included.

The next five chapters outline the basics of vegetation science that are necessary for understanding Part 2 of this book. Vegetation, or any other subject of study, can be approached from eight viewpoints that together form a cohesive whole. These viewpoints are:

Composition— Of what is it made?
Structure—   How are the components arranged?
Function—   What does it do?
Ecology—   How does it interact with its environment?
Development—How does it form?
Classification—How is it categorized?
Distribution— Where is it found?
History—   From where did it come?

2

COMPOSITION AND STRUCTURE

Of the eight viewpoints or approaches to the study of vegetation, composition and structure form a closely related pair—the first deals with the components of vegetation and the second with the arrangement of these components.

COMPOSITION

The essential part of any study of vegetation is the composition approach. Just as a question about the composition of a piece of cloth fabric may be answered with a response of "cotton and wool," the question of the composition of a type of vegetation may be at least partially satisfied with a listing of its components, that is, its species. With such a list the reader can begin to visualize the vegetation, provided he or she is familiar with the species. Unfortunately, the number of plant species in North America is so vast that it is impossible for even a trained taxonomist to know more than a fraction of them.

In order to avoid this problem with species lists, the growth-forms of the various species may be described, instead of using their individual taxonomic names. In the growth-form system, plants are classified according to their morphology and period of growth, regardless of taxonomic affinities. One set of growth-form terms that has been used at least as far back as the ancient Greeks is that of trees, shrubs, and herbs. The general meaning of these terms is well known, but the botanist has fairly rigorous definitions for them. A tree is a woody plant that typically has only one major stem. A shrub is also a woody plant, but one that usually has several main stems. A herb is a nonwoody plant and hence has a stem that is usually short-lived.

Another set of growth-form terms is based on the length of the period of growth. An annual is a herb that lives only a single growing season; both vegetative growth and reproduction take place in one season. A biennial is a herb that lives two growing seasons; vegetative growth predominates during the first, and the reproduction process is completed in the second. A perennial lives for at least several years, not dying after reproducing. Clearly, trees and shrubs are perennials but so are the many herbs whose above ground parts may die at the end of a growing season, but whose bulb or analagous structure persists below ground.

Some growth-form classifications can be much more detailed; yet, many terms such as evergreen, succulent, and broadleaf are already known by the nonbotanist and will be used throughout Part 2 of this book. Even though some terms may be unfamiliar, growth-form systems are much simpler than taxonomic classification, which uses individual species names. Growth-form systems are especially helpful in the study of taxonomically diverse vegetation, such as many tropical forests. A second benefit of the growth-form approach will become apparent when various kinds of vegetation are compared. Not only will differences in the presence and frequency of various growth-forms help in this comparison but, since growth-forms are thought to represent adaptations, they will assist the reader in recognizing relationships between vegetation and its environment.

Although much information may be presented in growth-form studies, there is significantly more information given in a list of the species found in a type of vegetation. This is assuming that the reader is familiar with the species on the list, that the species list is complete, and that the ecological requirements and relationships of most species are known.

Unfortunately, there is the previously noted problem of unfamiliarity with taxonomic names. Also, most studies do not result in species lists that are complete, that is, include all seed plants, ferns, mosses, lichens, algae, and so on. Even complete lists must be interpreted carefully. The presence of a species indicates that the environment is within its **tolerance limits**—the range of environmental conditions under which the species may grow. However, the absence of a species does not necessarily mean that the environment is outside of its tolerance limits. Instead, other factors may be involved, such as the lack of a nearby seed source. An additional problem with describing vegetation by species lists is that, with the possible exception of some agricultural species, our knowledge of the ecological role of even one species is limited at best.

Today, vegetation cannot be described by species lists alone. In fact, plant ecologists recognize a difference between species lists and descriptions of vegetation. A species list is known as a **flora,** an enumeration of the species

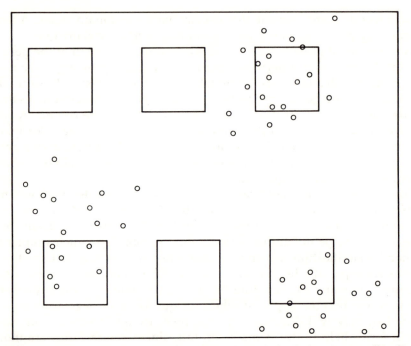

FIGURE 2.1   Diagram of a population with a relatively high density, small individuals, and moderate frequency, as sampled by meter-square quadrats (density = 3.7 individuals/m²; frequency = 50 percent).

present without regard for their relative importance or dominance. In contrast, the term vegetation, as defined in Chapter 1, implies such knowledge. And just as the answer of "cotton and wool" to the question of fabric composition is incomplete without an indication of the relative amounts of each, so is a flora an inadequate description of vegetation without an indication of the relative importance or dominance of its individual species.

Vegetation scientists find it difficult to define **importance** and **dominance.** Ideally, these terms should express the degree of a species influence on the ecosystem; however, this is not a practical definition because our knowledge of species ecological roles is so limited. Instead, plant ecologists are forced to use relatively easily measurable attributes of species such as density and size and assume that they correlate with ecological roles. Most researchers agree that any such attribute alone is insufficient as an indicator of importance or dominance and that several complementary factors should be used.

Density, the number of individuals per unit area, is often used under the assumption that the more numerous a species is within an area, the more likely that it is an important or dominant species. Even ignoring problems

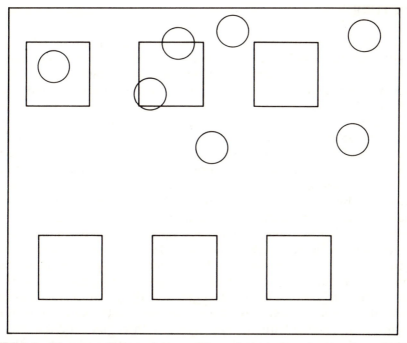

FIGURE 2.2   Diagram of a population with a relatively low density, large individuals, and low frequency, as sampled by meter-square quadrats (density = 0.5 individuals/m²; frequency = 33 percent).

associated with computing density (for instance, deciding whether an entire clump of prairie grass is one individual or if each stem is an individual), it should be clear that, since density values indicate nothing about size nor about how widespread a species is, density alone is not a sufficient definition of importance or dominance (Figure 2.1).

Size is a second factor commonly considered an element of importance and dominance. There are many aspects of size and, therefore, many ways to measure it. Some ecologists use basal area (stem area per unit area), others use cover (the ground area covered by a projection of plant crowns), and still others use biomass (usually taken as the weight of the dried organic matter). All the many methods used to measure size have advantages and disadvantages, depending on the type of vegetation being studied. But again, size alone cannot adequately define importance or dominance, since it does not indicate the density nor the distribution of the species (Figure 2.2). However, when size and density are used in combination, they lack only a measure of the species distribution. This deficiency is resolved in the structural approach to the study of vegetation.

STRUCTURE

Structure, like composition, has an essential role in the presentation of the different formations in Part 2 of this book. The structure of vegetation refers to the distribution or arrangement of species and growth-forms. First, the distribution of individuals of a single species will be discussed and, second, all individuals of all species making up a plant community will be studied.

The discussion of importance and dominance concluded that density and size need to be combined with a measure of how widespread a species is. The factor used to describe the extent of a species distribution is **frequency.** It is determined by establishing a series of small study plots (quadrats) of some standard size within a stand of vegetation and calculating the percentage in which the species occurs (Figure 2.3). It is assumed that the greater the frequency value, the more important or dominant the species is in that stand. However, note that frequency values are highly dependent on quadrat size.

Frequency interrelates with density and plant size (and dispersion, which is discussed next) in a complex way. The density, size, and frequency factors combined allow for an estimation of the importance or dominance of individual species in various kinds of vegetation (in place of the unavailable measure of a species influence on the ecosystem). Note that any one of these attributes alone is inadequate but, with all three in combination, each compensates for the major shortcomings of the other two. A method for calculating importance (dominance) using quantitative values of these three factors may be found in many general ecology textbooks.

The distribution of the individuals of a species also may be described from the standpoint of their pattern. Three general types are shown in Figure 2.4. One is the **random** pattern. It rarely occurs, because it is achieved only in the unlikely event that no individual has any influence on the location of any other individual. Occasionally, this pattern may be observed when a species first invades a new habitat, provided the site is uniform and there is random dispersal of the seeds. A second type of pattern, **regular,** is unusual in natural situations, but common in artificially maintained communities such as orchards. Desert shrubs occasionally have a regular distribution. Their pattern, which is usually not regular though the individuals are widely spaced, is thought to be maintained by root competition for water and/or by a chemical factor (see page 186).

Most plant species are arranged in a third pattern, **clumped,** because habitats typically are not uniform, and because individual plants usually have an irregular influence on the location of other individuals. For example, as plants change the microclimate and soil conditions around themselves (through casting shade and other means), they produce small-scale habitat variation. Plants of other species will tend to occur in the sites most suited to their requirements, be it beneath or away from the individuals of the first species, and a clumped pattern will develop. This pattern also arises when a

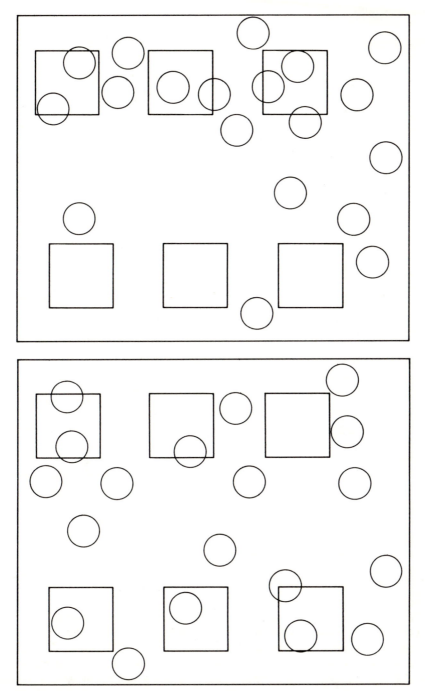

FIGURE 2.3 Diagrams of two populations with equal densities, equal size individuals, and different frequencies, as sampled by meter-square quadrats [densities = 1.2 individuals/m²; frequency (top) = 50 percent; frequency (bottom) = 83 percent].

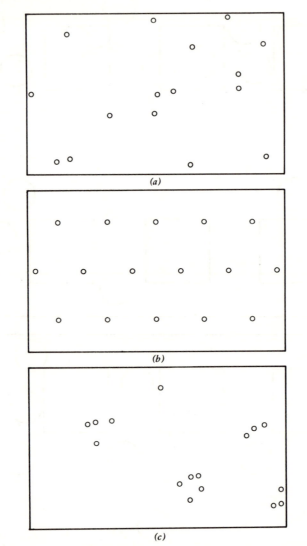

FIGURE 2.4 Diagrams of (*a*) random, (*b*) regular, and (*c*) clumped dispersion patterns.

plant species reproduces vegetatively or disperses seeds ineffectively, and as a result the young plants are grouped around the parent plants.

Up to this point, the discussion of structure has dealt with the distribution of individuals of single species. But in considering an entire community, the most important aspect of structure is that of **stratification**—the vertical arrangement of plants into a few layers instead of an even distribution throughout all heights from the ground to the tallest plants (Figure 2.5).

FIGURE 2.5   Diagrams of vertical stratification in a mature deciduous forest (top) and a mature Tall Grass prairie.

Stratification is more apparent in some forests than in other types of vegetation. In a typical mature deciduous forest, one may discern a surface layer (frequently of mosses and lichens), a herb layer, a shrub layer, a subcanopy tree layer, and a canopy tree layer (note the relationship between vertical

FIGURE 2.6   Diagram showing the lack of vertical stratification in a young deciduous forest.

structure and growth-forms). It is generally possible to observe some vertical structure in any type of mature vegetation; however, layers may not be well defined in young stands (Figure 2.6). Unfortunately, the ecological significance of stratification is clouded by the fact that any single layer may include species with diverse requirements. Nevertheless, descriptions of vertical structure are useful in visualizing some of the gross differences between various kinds of vegetation.

3

FUNCTION

The "function" of vegetation refers to "what vegetation does." A list of functions could be nearly endless, but this chapter concentrates on the role of vegetation as related to the flows of energy and matter in ecosystems. The community, including the plants which form the vegetation, makes up the biotic component of ecosystems. The abiotic component refers to the nonliving parts of ecosystems; the nonliving parts are divided into matter and energy. Water and minerals are examples of matter. Solar radiation is a form of energy. As matter and energy flow in an ecosystem, they are changed in form numerous times, although the total amount of either remains constant. Matter is never changed into a totally nonreusable form, so it has a cyclic pattern of flow in ecosystems. In contrast, following the second law of thermodynamics, energy transformations always result in some energy being degraded into heat, a form of energy that is not reusable. Energy, therefore, exhibits a noncyclic pattern of flow, and ecosystems are dependent on a continued input of energy.

ENERGY FLOW

The flow of energy through vegetation and its ecosystem begins with incoming solar radiation. Only about 57 percent of the solar radiation which hits the atmosphere actually reaches the earth's surface; the rest is reflected back to space or is absorbed by clouds. The distribution of that light which hits the earth's surface is highly variable both in time and location. In addition to such obvious factors as cloud cover, the amount of incoming light is influenced by latitude and even topographic relief, as will be discussed in the following chapter. Regardless, after input of light, the flow of energy follows what is termed a **food chain.** A familiar example of a food chain is the grass-cattle-human system in which energy flows from one organism to the next as individuals are eating or being eaten. The term "chain" is used because the three organisms are linked together in a chain of energy relationships.

Although the food chain concept is useful, it is oversimplified; in fact, all species are involved in numerous interconnecting food chains that collectively form a **food web** (Figure 3.1). The species that make up a food web may be categorized by their role in the network. These categories are known as **trophic levels;** they indicate the point in the food web where species receive energy, that is, where they "feed." The base of a food web is the **producer** trophic level; this level includes all the green plants—the organisms that produce food by converting the incoming energy of solar radiation through the process of photosynthesis into the chemical energy of organic compounds.

Despite the adaptations of plants which aid their abilities to absorb solar radiation, surprisingly little of that which hits the earth's surface is actually used in photosynthesis. The total amount of chemical energy produced in photosynthesis is known as **gross primary production.** Of course, this amount is variable; for example, a deciduous forest in full leaf is obviously a more efficient solar collector than desert vegetation (Table 3.1). Worldwide, the percent of annual surface solar radiation used in gross primary production averages approximately 0.11 percent (0.06 percent for the oceans and 0.24 percent for the land surfaces). In the most productive stands of natural vegetation, daily gross primary production rarely exceeds 3 percent of solar radiation. Intensive cultivation may achieve a figure of 6 to 8 percent.

Gross primary production is used by plants in a variety of ways involving their growth, reproduction, and maintenance. For example, a large portion goes to respiration, a process that changes the stored energy of the organic compounds into other, more immediately useful forms of chemical energy and into heat energy, which is released (and lost to the ecosystem). The remaining portion of the photosynthetic output is, in effect, stored by the plant as the cellulose of cell walls, the proteins of intracellular bodies, the starch of storage cells, and so on. This portion of the gross primary production, which remains after respiration, is termed **net primary production.** Estimated annual values for a variety of ecosystem types are given in Table 3.1.

The net primary production is available to (but not entirely used by) the food web's next trophic level, the **primary consumer** (or herbivore) level. The species of this trophic level are adapted to a plant diet; the organic compounds they assimilate are used in their growth, reproduction, and maintenance. Once more, respiration results in a loss of heat energy from the ecosystem. The energy that remains is converted into new tissue and cellular storage compounds. This is what is available to the next trophic level, the **secondary consumer** (or primary carnivore) level. Similar processes take

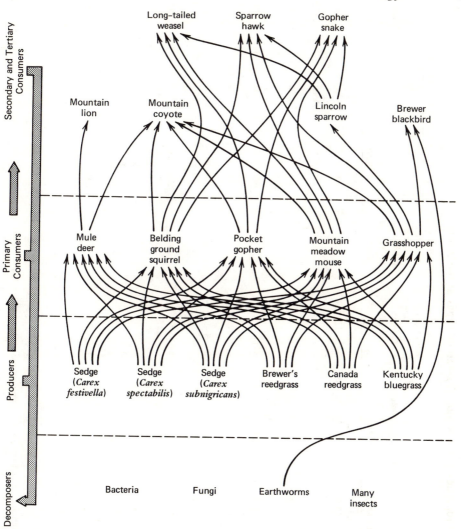

Secondary and Tertiary Consumers

Primary Consumers

Producers

Decomposers

Long–tailed weasel

Sparrow hawk

Gopher snake

Mountain lion

Mountain coyote

Lincoln sparrow

Brewer blackbird

Mule deer

Belding ground squirrel

Pocket gopher

Mountain meadow mouse

Grasshopper

Sedge (*Carex festivella*)

Sedge (*Carex spectabilis*)

Sedge (*Carex subnigricans*)

Brewer's reedgrass

Canada reedgrass

Kentucky bluegrass

Bacteria

Fungi

Earthworms

Many insects

FIGURE 3.1 Diagram of part of the food web of a mountain meadow ecosystem (overlapping with surrounding communities) in the Sierra Nevada mountains of California. To avoid confusing detail, the flow of energy from individual species to the decomposers is not shown.

place in this level and result in the loss of heat energy and in the availability of animal tissue (stored energy) to a possible fourth trophic level, the **tertiary consumer** (or secondary carnivore) level. Tertiary consumers are not present in all ecosystems, because energy losses lower in the food web are sometimes so great that they leave an amount of energy insufficient to support this level.

Organisms such as bacteria and fungi are the **decomposers** of an ecosys-

TABLE 3.1 Energy-Related Properties of the Major Vegetation Types of the World

Vegetation Type	Area (10⁶ km²)	Range of Maximum Biomass of Mature Communities (kg/m²)	Range of Leaf Surface Area to Ground Surface Area Ratio (m²/m²)	Efficiency of Radiation Utilization, Average for the Year in Percent	Net-Primary Productivity Range (g/m²/year)	Net-Primary Productivity Approximate Mean	Total Production (10⁹ t)
Forest	50.0					1290	64.5
Tropical rain forest	17.0	45, 75?	6-10-12-16.6	4.5	1000-3500	2000	34.0
Raingreen (tropical seasonal) forest	7.5	42	6-7-10		600-3500	1500	11.3
Summergreen (temperate deciduous) forest	7.0	42-46	3-12	1.6	400-2500	1000	7.0
Chaparral	.15	26	4-7-12	1.0[a]	250-1500	800	1.2
Warm temperate mixed forest	5.0	24	5-14		600-2500	1000	5.0
Boreal forest	12.0	20-52	7-15	1.1	200-1500	500	6.0
Woodland	7.0	2-20	4.2		200-1000	600	4.2
Dwarf and open scrub	26.0					90	2.4
Tundra	8.0	0.1-3	0.5-1-1.3	0.6	100-400	140	1.1
Desert scrub	18.0	0.1-4	?	0.05[b]	10-250	70	1.3
Grassland	24.0					600	15.0
Tropical grassland	15.0	?-5	1-5	0.6	200-2000	700	10.5
Temperate grassland	9.0	?-3	?-5-9-16	0.6	100-1500	500	4.5
Desert (extreme)	24.0					1	—
Dry desert	8.5	0	0		0-10	3	—
Ice desert	15.5	0	0		0-1	0	—
Cultivated Land (annual crops)	14.0	3.5	4-12	0.7	100-4000	650	9.1
Freshwater	4.0					1250	5.0
Swamp and marsh	2.0	2.5-?	?-11-23.3		800-4000	2000	4.0
Lake and stream	2.0	?-0.1	?		100-1500	500	1.0
Total for continents	149.0					669	100.2

[a] For sclerophyllous woodland vegetation.
[b] For semidesert vegetation.

Sources. Lieth, H. 1975. Primary production of the major vegetation units of the world, p. 203–215. *In* H. Lieth and R. H. Whittaker (eds.). Primary productivity of the biosphere. Springer-Verlag, New York. Larcher, W. 1975. Physiological plant ecology. Translated by M. A. Biederman-Thorson. Springer-Verlag, Berlin. 252 p.

tem. They "feed" directly on all other trophic levels by receiving energy from the digestion and respiration of the organic compounds in dead leaves, stems, animal tissues, feces, and other such organic matter. Their respiration results in an additional loss of heat energy.

In a balanced ecosystem the respiration of all organisms produces a total heat loss equal to the gross primary production, that is, the amount of energy put into the ecosystem by the producers. In young developing ecosystems, gross primary production exceeds community respiration, so there is an excess of organic matter that accumulates as "storage" (Chapter 5). However, even this stored energy is eventually converted to heat, resulting in energy having a noncyclic pattern of flow: energy enters an ecosystem in the form of solar radiation, is changed into chemical energy by the producers, and is passed along the food web being degraded at every trophic level into heat energy, which is lost to space. Thus, ecosystems are dependent on continued inputs of solar radiation.

Figure 3.2 presents a generalized diagram of the flow of energy through an ecosystem. It illustrates the point that, though people tend to think in terms of food chains such as the grass-cattle-human chain mentioned previously, terrestrial ecosystems have greater energy flow through the decomposers than the higher consumers. Food chains with this characteristic are known as **detritus food chains.** They are also important in shallow-water ecosystems such as marshes. Food chains with greater energy flow through the consumers are known as **grazing food chains.** Despite their prevalence in lectures and textbooks, they are predominant only in deep-water ecosystems such as open oceans.

MATTER FLOW

The movement of organic compounds along the food web also can be used to illustrate the flow of matter. In energy flow, respiration eventually results in the change of energy into a biologically unusable form (heat), but matter flow never involves changes into states that are not, at least potentially, usable by other organisms. Therefore, matter has a cyclic pattern of flow, even though individual atoms may take eons to be "recycled." The chief role of vegetation in most matter cycles is that it converts chemicals into biologically useful forms, that is, forms that can continue the flow along the food web. The cycles of matter are known as **biogeochemical cycles** (bio refers to "life," geo refers to "earth, soil, and air," and chemical refers to the elements or compounds involved). The cycles may be divided into two categories: gaseous and sedimentary. In **gaseous cycles** there is a major storage of the element in a gas phase, usually in the atmosphere. In **sedimentary cycles** the chief site of storage is in rocks, the weathering of which makes the element

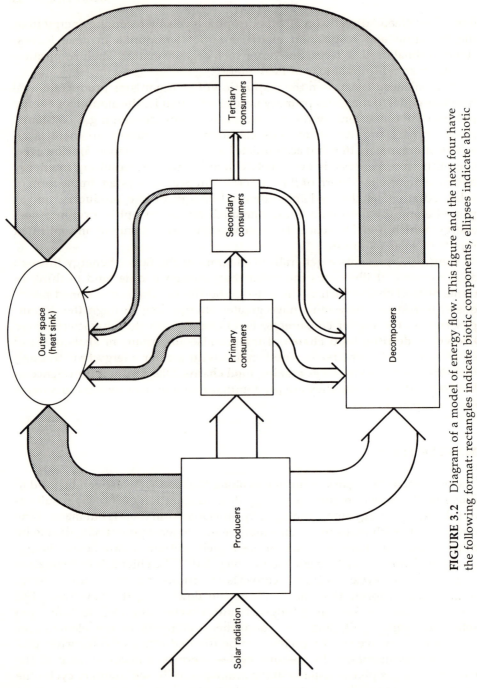

FIGURE 3.2 Diagram of a model of energy flow. This figure and the next four have the following format: rectangles indicate biotic components, ellipses indicate abiotic components, and the connecting arrows indicate flows.

available to organisms. First to be considered are the gaseous cycles of carbon, oxygen, and nitrogen.

The connection between matter and energy flow is best illustrated by carbon, the basic structural element of all organic compounds. These compounds are, of course, initially produced in the process of photosynthesis, which uses carbon in the gaseous form of carbon dioxide. The energy, carbon, and other chemical elements of the organic compounds are passed along the food web, with carbon being released as carbon dioxide whenever respiration occurs. The reuse of this carbon dioxide in photosynthesis results in a cyclic flow (Figure 3.3). Carbon dioxide and other compounds are also released in the combustion of wood, coal, oil, and natural gas, all of which are composed of organic compounds that are produced by plants.

Photosynthesis is also a key process in the oxygen cycle. Atmospheric oxygen is produced in photosynthesis; however, the role of vegetation is not limited to this. Plants remove oxygen from the atmosphere as it is used in their respiration. The oxygen and organic compounds produced in photosynthesis also are used by other organisms in their respiration.

A third gaseous cycle is that of nitrogen (Figure 3.4). This element is a component of all amino acids, the compounds which make up proteins, and proteins are the chief structural material of all organisms. Eighty percent of the earth's atmosphere is molecular nitrogen gas, a chemical form that is not directly usable by plants. It becomes available for plant growth primarily through **biological nitrogen fixation,** a process carried on by certain species of blue-green algae and bacteria. Some nitrogen-fixing bacteria are free-living in the soil, but most are symbiotic with plants and are in nodules (sphere-shaped structures) on the roots of certain species, such as some

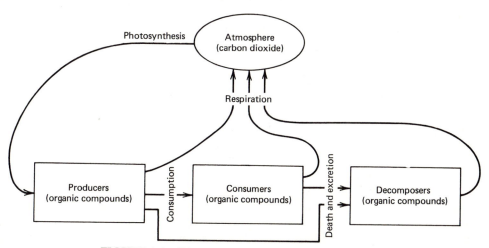

FIGURE 3.3 Diagram of a generalized carbon cycle.

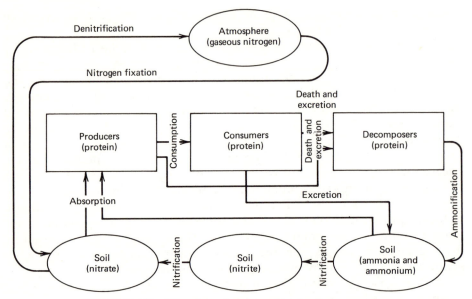

FIGURE 3.4   Diagram of a generalized nitrogen cycle.

legumes. Not all legumes have the bacteria and, hence, the capacity to convert atmospheric nitrogen into a biologically usable form, but soybeans, alfalfa, and clover are familiar examples of plants with this capability. Some nonleguminous plants also have nodules of nitrogen-fixing bacteria. These bacteria have the plant's photosynthetic products as their food source and change nitrogen gas into nitrate, which is used by the plant in the production of amino acids.

The amino acids are passed along the food web and are used in the production of protein at all trophic levels. Eventually the decomposition of plant and animal tissues and waste products changes the nitrogen of the amino acids into ammonium through the process of ammonification. Some ammonium is absorbed by certain plant species and is used in the production of new amino acid molecules, but much of it remains in the soil and is changed into nitrite. This process is known as nitrification. It is carried on by a certain group of bacteria (the *Nitrosomonas* genus), which obtains its energy from the conversion of ammonium to nitrite. A second group of bacteria (the *Nitrobacter* genus) carries the nitrification process a second step and, in obtaining its energy, changes the nitrite ions to nitrate. The nitrate may be absorbed and used by plants to produce amino acids, but some is changed into atmospheric molecular nitrogen by the denitrification process of certain bacteria and fungi. Such losses from the biological system are balanced by a variety of inputs, especially from biological nitrogen fixation.

Elements with a sedimentary biogeochemical cycle do not have a large gaseous reserve. Some elements, such as sulfur, have a gas phase (sulfur dioxide) but, since supplies in the earth's crust are more important, their cycles are still considered to be of the sedimentary type. Phosphorous has an obvious sedimentary cycle because it basically lacks a gas phase. It is an essential element—a component of the genetic code molecule, DNA, and of the cellular energy "currency" compound, ATP. The amount of phosphorous in a biologically usable form is usually low in the environment. The sudden growth of masses of algae in lakes and rivers following the input of waste water containing phosphorous indicates that the natural low supply of this element limits growth in many aquatic ecosystems.

Phosphorous enters the biotic component of the ecosystem when it is absorbed in the inorganic form by plant roots (Figure 3.5). It is converted into organic phosphorous and passed along the food web (with other elements and energy). The decomposition of tissues and wastes mineralizes the element, returning it to the soil where plants may reabsorb it in the inorganic form. However, some soil phosphorous is continually lost from this cycle. It is dissolved on and in the soil by precipitation and removed by the flow of surface and ground water to large bodies of water such as lakes and oceans.

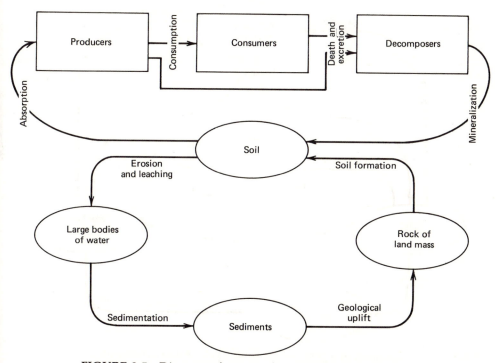

FIGURE 3.5 Diagram of a generalized phosphorous cycle.

Here the phosphorous may be used by aquatic organisms (the cycles previously discussed in this chapter are also found in aquatic ecosystems), but phosphorous in sediment is continually deposited on the lake or ocean floor. Unless the sediments are brought near the surface by water upwelling, this deposition removes phosphorous from the biotic component; yet, over eons of time these materials are recycled when geological forces uplift new land masses and rock formed from the sediment is weathered. This returns the minerals to a state where they are again available for root absorption and entrance into the food web.

The next cycle to be considered is that of water—a chemical compound essential for life. Water is of great importance in biogeochemistry, since many elements are dissolved in it throughout most of their cycles. Also, the distribution of many general vegetation types is correlated with the distribution of precipitation. Over 90 percent of the water on earth is in the oceans, and most of the freshwater is tied up in glaciers and permanent snow. Perhaps only 1 percent occurs in lakes, rivers, the atmosphere, ground water, soil, and organisms. The water cycle involves the flows between all of these (Figure 3.6).

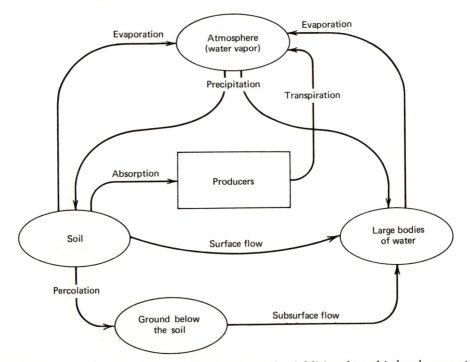

FIGURE 3.6 Diagram of a generalized water cycle. Additional trophic levels are not shown because relatively little water passes along a food web.

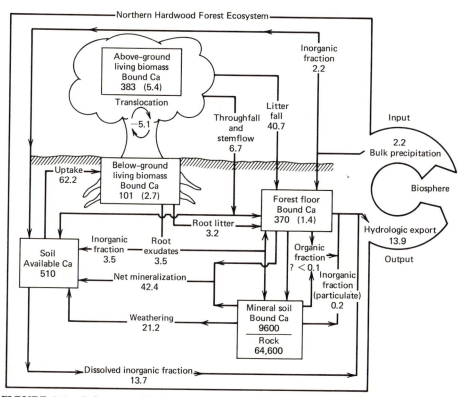

FIGURE 3.7 Calcium cycle for a young, developing forest ecosystem at Hubbard Brook, New Hampshire. Values of components are in kilograms of calcium per hectare; values of flows are in kilograms of calcium per hectare per year. Values in parentheses are annual accretion rates of calcium. *Source.* Likens et al. 1977. Biogeochemistry of a forested ecosystem. Springer-Verlag, New York. 146 p.

Vegetation plays several roles in the water cycle. **Transpiration** is the evaporation of water from organisms, mostly from plants. It adds water vapor to the atmosphere and results in reduced soil moisture. Another effect of vegetation is that a plant canopy intercepts precipitation. This reduces surface runoff, because some of the intercepted water evaporates and the rest slowly drips or flows to the ground where a higher percentage will infiltrate than would if it came as intense direct precipitation.

The carbon, oxygen, nitrogen, phosphorous, and water cycles have been presented in generalized outline form. Detailed studies using actual examples of biogeochemical cycles in ecosystems are relatively recent and few in number. Figure 3.7 presents the calcium cycle for a young, developing forest in New Hampshire. Calcium has a sedimentary cycle. The detail shown in this figure is characteristic of all biogeochemical cycles.

4

ECOLOGY

The term ecology has been widely used in recent years as an all-inclusive term. In this book, however, ecology is only one of eight approaches to the study of vegetation. Nevertheless, since it refers to the interrelationships of vegetation and environment, all the other approaches can be put in an ecological context. For example, structure can be related to environment, development can be related to environment, and so on. In fact, vegetation scientists usually use an ecological approach when interpreting various aspects of vegetation.

The clearest method for discussing the ecology of vegetation is to list the major environmental factors that affect it and explain them one by one. This chapter will consider the climate, soil, topographic relief, biota, and fire factors. The danger in this approach is that the reader will not fully recognize the extent to which environmental factors are interrelated. The concept that the environment is an interconnected whole is represented in the **holistic** viewpoint; an understanding of this viewpoint is essential for interpreting vegetation.

One of the implications of the holistic viewpoint is that it is difficult to determine the cause of many aspects of vegetation. Scientific study usually reveals correlations between vegetation and particular environmental factors, but correlation does not mean, nor even imply, causation. For example, there are great differences in vegetation from the base to the top of a mountain. The variation in vegetation can be correlated with differences in average temperatures; however, this does not mean that temperature has caused the variation in vegetation, nor that temperature is the only environmental factor involved. Indeed, differences in other climatic factors such as precipitation and wind, as well as soil factors, slope direction, and fire frequency also may be important. Only extensive study of individual vegetation differences could reveal the key environmental factor or factors involved in each case. Thus the problems of relating vegetation to environment are extremely complicated and pose an intriguing challenge for vegetation scientists.

CLIMATE

Although environmental factors are interrelated, climate is the one thought to be most important on a regional scale. Climate is the average weather of a particular location. It has many effects on vegetation: solar radiation is essential for photosynthesis, precipitation supplies soil moisture necessary for growth and maintenance, frosts limit the growing season, winds increase transpiration, and so on.

Climates obviously differ greatly over North America. Much of this variation correlates with differences in vegetation, so it is of primary importance to understand the major factors determining climate. Regular seasonal climatic changes and differences between locations are related to the earth's sphere shape, its 365-day orbit around the sun, and the 23½° angle of the earth's axis to the plane of this orbit (Figure 4.1). Differences in orbital

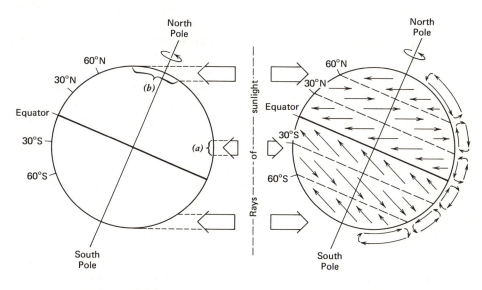

Summer Solstice   Winter Solstice

FIGURE 4.1 Diagram of the orbital position of the earth relative to the sun on about June 22 (left) and December 22 (right). The letters (a) and (b) on the left drawing illustrate why low latitudes receive more solar radiation per unit area than high latitudes. The right drawing shows air-circulation patterns.

distance from the sun are an insignificant factor. During the summer the northern hemisphere directly faces the sun, that is, the sun's radiation hits this portion of the earth at a high angle (left globe in Figure 4.1). The sun's rays are at a lower angle during winter (right globe). Fall and spring occur when the earth has intermediate positions relative to the sun. The seasons are reversed in the southern hemisphere.

The angle of incidence of solar radiation also has a great influence on temperature variations between different latitudes. The sun's rays hit the earth almost perpendicularly at low latitudes, but at high latitudes the angle is always much lower. As a result, the lower the latitude, the greater the radiation received per unit area (compare *a* and *b* on the left globe in Figure 4.1).

Different latitudes also have different day lengths, and these change continuously as the earth's orientation to the sun changes. Parts of the world along the equator show relatively little day-length variation; all are close to 12 hours in length. Variation increases away from the equatorial regions. Day lengths in the midlatitudes may be around 15 hours in the summer and only 9 hours in the winter. Toward the poles, summer days lengthen, and some have 24 hours of sunshine while the sun never rises on an equal number of winter days. The latitudinal and seasonal differences in day-night lengths are of great importance in understanding the distribution of various plant species.

Latitudinal temperature differences produce air-circulation patterns in the atmosphere (right globe in Figure 4.1). The equatorial region, because of the sun's nearly constant high angle, averages more solar radiation and more consistent high temperatures than other latitudes. This favors evapotranspiration (evaporation plus transpiration) and produces high humidities. Since warm air is lighter than cool air, the heated tropical air rises in the atmosphere. As this humid air cools, its capacity for holding water vapor decreases, and the moisture condenses and falls as precipitation. With exceptions, the equatorial region therefore tends to have high amounts of precipitation, as well as high average temperatures. Associated with this climate are various types of tropical vegetation, the most familiar being the tropical rain forest.

The climate at latitudes 30° north and south is dominated by high-pressure systems, areas of descending air. This air is dry, and as it approaches the earth's surface it warms, further reducing its low relative humidity. Consequently, with exceptions, there tends to be relatively low precipitation and high potential evapotranspiration at this latitude. Most of the world's deserts are located here. The 60° north and south latitude bands are regions of ascending air, and the poles are areas of descending air.

These air-circulation patterns also effect winds. Air flows from latitudes of high pressure to latitudes of low pressure, but these movements are modified by a coriolis effect—a deflecting force produced by the spinning of the earth on its axis. The net result is shown on the right globe in Figure 4.1. The wind pattern is from the northeast (the Trade Winds) between the equator and 30° north, from the southwest (Westerlies) between 30° and 60° north, and from the northeast (Easterlies) between 60° and the North Pole. Wind direction is especially important in mountainous areas, as discussed in the next paragraph.

There are many deviations from the general, global climatic pattern described above. For example, the coastline of the United States along the Gulf of Mexico is not a desert, despite its proximity to the latitude of 30° north. Three major factors that alter the general, global climatic pattern are tall mountain ranges, large bodies of water, and large land masses. The major regional effect of tall mountains is that they produce **rain shadows,** areas of very little precipitation. An air mass moving into mountains rises as it flows across the landscape. The air cools, moisture condenses, and precipitation falls on the windward side of the mountains. The other side has descending dry air, so there is little precipitation there and for some distance from the mountains. Since the wind pattern is from the west in midlatitude North America, rain shadows occur east of the Sierra Nevada Mountains (in the Great Basin) and east of the Rocky Mountains (in the Great Plains).

A second important factor influencing climatic patterns is the presence of large bodies of water. Oceans and large lakes produce vast amounts of water vapor that eventually fall as precipitation. They also have moderating effects on temperatures, because a physical characteristic of water allows it to retain large amounts of heat energy. Land masses obviously do not have this capacity. In fact, a major effect of land masses on global climates is the increasing of temperature extremes. Thus, the **maritime climate** of coastal areas is generally much more moderate than that of inland areas, which have what is known as a **continental climate.** For example, San Francisco does not have the extreme summer highs and winter lows that are typical of Chicago. San Francisco also has a more moderate climate than east coast cities, such as New York, in part because of the prevailing wind pattern (Westerlies).

The interaction of the various factors affecting the global climate results in precipitation and temperature patterns as shown in Figures 4.2 and 4.3. Note a similarity between portions of these maps and the distribution of the formations of North America (see the frontispiece). However, recall that correlation does not imply causation. Also, the discrepancies between these maps and the vegetation map serve as a reminder that environmental factors act holistically.

Many systems for classifying climates have been proposed. Frequently, these indicate a close relationship between climate and vegetation. In fact, such words as desert and tropical are often applied to both climate and vegetation. Climates have sometimes been classified, in part, by using

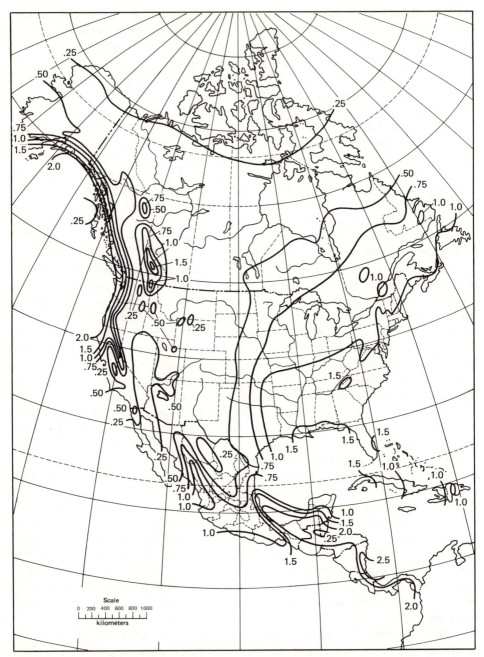

FIGURE 4.2 Precipitation patterns across North America. Lines connect locations with approximately equal total annual precipitation. Values are in meters (1.000 m = 1000 mm).

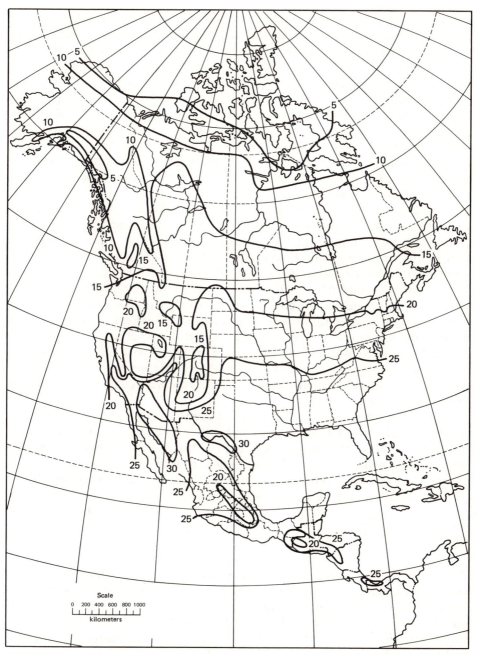

FIGURE 4.3 Temperature patterns across North America. Lines connect locations with equal, July-mean temperatures (° C).

vegetation as an indicator. In those instances, there is an automatic and meaningless correlation between the distribution of vegetation and climate.

No classification of climates is used in this book. Instead, Part 2 will include diagrams of climates thought to be typical of particular formations. Examples of these diagrams are given in Figure 4.4. When studying these graphs, keep in mind that they illustrate regional climates. They do not indicate **microclimates,** which are local variations in the regional climate produced by such factors as topographic relief and vegetation cover. The climate diagrams also do not depict rare and extreme droughts, frosts, winds, and such things. Yet both microclimate variation and unusual extreme weather are very important in the distribution of individual species, communities, and even formations.

At the top of each climate diagram is the location of the weather station, its elevation (in parentheses, in meters), the average annual precipitation (mm), and the average annual temperature (° C). Each diagram is a graph of two factors: mean monthly temperature (° C) and mean monthly precipitation (mm) for January through December. The scale of the vertical axis is in 10° C—20 mm subdivisions. Those months with the precipitation line above the temperature line may be considered relatively wet (because precipitation should exceed evaporation). This is indicated by vertical lines; solid black areas at the top of the lined areas indicate that the precipitation scale above 100 mm has been reduced by 90 percent to keep the graph at reasonable height. If the temperature line is above the precipitation line, the months are considered relatively dry (because of high potential evaporation) and are indicated by dots.

Below the horizontal axis of most diagrams, various months have solid, lined, or clear areas. The solid areas indicate months with a mean daily minimum temperature at or below freezing. The lined areas are beneath months with a higher mean and at least a single day with below-freezing temperature. The clear areas represent months without any freezing temperatures. This information may be lacking for some locations.

SOIL

A second of the important environmental factors that interact to affect vegetation is soil—a complex and dynamic part of terrestrial ecosystems. The definition of soil as the medium for plant growth illustrates its strong relationship with vegetation. The components of soil are various sized rock fragments, decaying organic matter, living organisms, atmospheric gases, and water solutions.

The rock fragments are remnants of the original material, the **parent material,** which weathered to form this matrix of the soil. The names for the

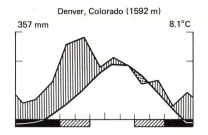

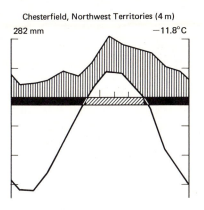

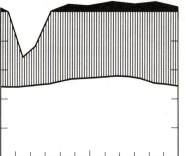

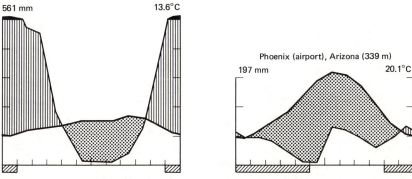

FIGURE 4.4 Examples of climate diagrams. *Source*. Redrawn from Walter, H. and H. Lieth. 1967. Klimadiagramm-weltatlas. Fischer, Jena, East Germany.

different sized fragments, beginning with the largest, are gravel, sand, silt, and clay. The proportions of the last three determine **soil texture.** Soils with a balance of these three rock fragments are referred to as loams. Other soils are named on the basis of the dominant rock fragment size: silt-clay, clay-loam, sandy-loam, and so on. Although a trained soil scientist can estimate soil texture from feel, laboratory methods determine it more precisely.

Sand and clay particles are especially important in soils. Sand grains, because of their size, have relatively large pore spaces between them. This space aids the movement of soil gases, solutions, and even organisms. Yet soils that are nearly all sand are usually poor for plant growth; they are too well drained and frequently infertile. Clay particles, on the other hand, are tightly packed because of their small size. Their minute pore spaces act as billions of capillary tubes to hold moisture in the soil. The small clay particles also have great exposure of surface area to which minerals may bond and be retained in the soil (instead of being dissolved and removed by infiltrating precipitation). Soils that are nearly pure clay, however, are also relatively poor for plant growth; the small pore spaces retard the movement of soil gases, solutions, and plant roots. But when one considers the advantages of sand and clay particles combined, it should be apparent why loam soils are generally agriculturally productive.

A second component of soils is decaying organic matter, much of which is the remains of dead plant parts such as leaves, but earthworms and other soil organisms are also important contributors. Organic matter is the source of energy for decomposers. During decomposition, nutrients are mineralized and become available for absorption by roots; hence, organic matter is obviously important to soil fertility. Organic matter also increases the water-holding capacity of a soil. Furthermore, it is responsible for soil structure, sticking rock fragments together into clusters and thus increasing pore space. Finally, the color of a soil is largely determined by the amount of organic matter it contains; the more organic matter, the darker the soil, provided other factors are the same.

Living organisms are a third soil component. Included in this category are earthworms, bacteria, moles, insects, fungi, plant roots, and many others. Obviously, they have a range of individual roles; some of their major functions include the addition of organic matter, the increasing of pore space, the mixing of soil, the decomposition of organic matter, the fixation of nitrogen, and the absorption of water and minerals.

A fourth component is the soil atmosphere, that is, the below-ground gases. It fills the pore space that is not taken up by the soil solution. The importance of this atmosphere is illustrated by the death of many plants in waterlogged soil. Soil oxygen is necessary for the respiration of root cells in plants, as well as the cells of other oxygen-consuming organisms. Nitrogen gas, of course, is used by nitrogen-fixing organisms.

The last component is the soil solution. It is made up of the soil water and materials dissolved in it. The amount of solution in a soil can never be greater than the amount of pore space. The soil solution is the source of both water and minerals for plant growth and, hence, has a great influence on species' distributions.

The soil components take decades, if not centuries, to fully integrate into a mature soil. The major factors that influence the soil-formation process are climate, relief, fire, biota (including the vegetation), and parent material. As mentioned before, parent material is that matter which breaks down to form rock fragments—the chief substance of soil. The parent material may consist of solid bedrock or of material transported to the site by wind, water, or glaciers.

Weathering of parent material involves physical, biological, and chemical factors. Physical weathering includes the splitting of rocks through the repeated freezing and thawing of water in cracks. Biological weathering can accomplish the same thing when roots grow into cracks. Chemical weathering occurs when rock minerals are changed to different forms because of exposure to compounds such as water and molecular oxygen. Soil formation also involves the addition of living organisms and organic matter, the transport of various substances by infiltrating precipitation, and other activities.

Eventually, a **soil profile** develops. It consists of different layers or **horizons** that are chemically, physically, and, therefore, often visually distinctive (Figure 4.5). Soil profiles can be studied by digging a narrow trench to get a "profile" view of the different soil layers. The four major layers of a soil profile are the O, A, B, and C horizons. Any of these may be subdivided and labeled, for example, as B1, B2, B3, and so forth. The O (for organic) horizon consists primarily of decaying organic matter, such as dead leaves and feces, and the decomposers. Partially decomposed organic matter is carried into the A horizon by infiltrating precipitation. This water also may dissolve various minerals and carry them and clay particles down through the A horizon; this process is known as **leaching.** Leached materials are frequently deposited in the B horizon. The C horizon consists of relatively unweathered parent material.

A major environmental factor affecting the formation of soils is the ratio of precipitation to evaporation. Based on this ratio, there are two major soil groups—pedalfers and pedocals. **Pedalfer** soils develop in areas with a wet climate, that is, where precipitation exceeds evaporation. Consequently, they are highly leached and acidic. Under cool temperatures they are formed by a process known as **podzolization,** which involves the leaching of soluble minerals (especially bases) and colloids from the A horizon. The leaching is produced by the excess water interacting with the acid decomposition products of the leaf litter. Podzolization usually occurs beneath coniferous forests, where the leaf litter is always highly acidic. Under warm temperatures (with different vegetation) pedalfer soils are formed by **laterization,** a process

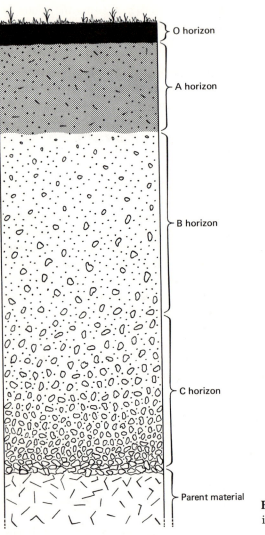

FIGURE 4.5 Diagram of a generalized soil profile.

involving intense leaching. Silica and bases are leached from the soil, and iron and aluminum are left as the major elements. Pedalfer soils formed by laterization are mildly acidic.

Pedocal soils are formed by **calcification**—a process that occurs under dry conditions where potential evaporation is greater than precipitation. Pedocals, therefore, are not well leached. They tend to be basic soils and are characterized by a calcium carbonate layer at the depth to which precipitation leaches calcium (0 to 2 meters, depending on local precipitation). The layer of calcium carbonate can become quite dense, forming a "hard pan" that inhibits root penetration.

The podzolization, laterization, and calcification processes all occur in their respective climates on well drained soils. In poorly drained areas, for example in much of the arctic tundra, a process known as **gleization** occurs. Water in the soil produces grey or bluish colored horizons because of the chemical reduction of iron into ferrous compounds. Gleization frequently produces soils with dense horizons lacking much structure. The presence and activity of decomposer microorganisms is low, so soils without good drainage usually have high amounts of organic matter.

The features of mature soil profiles vary greatly, depending on the nature of the environment. This variation provides means for more detailed soil classifications. One intricate, technical, but usable system is the "7th Approximation" classification of the United States Department of Agriculture. It is especially useful for detailed classification of local soil variations. A good introduction to this system is presented in the book by D. Steila, which is cited in the Suggested Readings for Further Study, following Chapter 6. For our purposes, the more general, international system originally developed in Russia is sufficient. It uses for categories those soils that develop on areas of level terrain or low relief and that correspond to mature vegetation types (hence, there is an inherent correlation with vegetation). These major soil types are known as the **Great Soil Groups.** Figure 4.6 is a diagram of the distribution of the Great Soil Groups of North America in relation to temper-

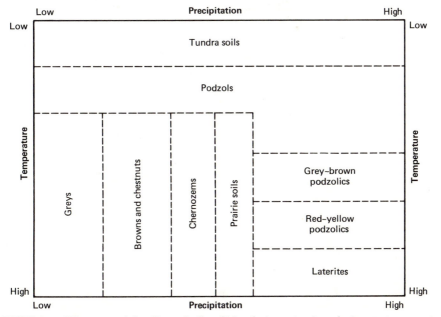

FIGURE 4.6   Diagram of the Great Soils of North America in relation to temperature and precipitation gradients.

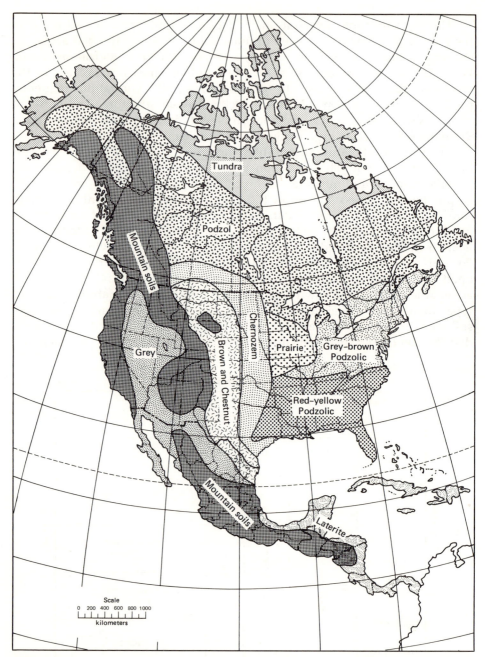

FIGURE 4.7 Map of the Great Soils of North America. Areas of mountainous terrain have a wide variety of soils but are here collectively shown as mountain soils. Many of these are rocky and have a poorly developed profile.

ature and precipitation gradients. Since these gradients are also the major climatic trends east to west and north to south across the continent, this chart resembles the Great Soils distribution map (Figure 4.7).

Tundra soils are variable but are generally shallow, dark in color, high in organic matter, subject to frost action, and possess permafrost (a permanently frozen layer). They support tundra vegetation. **Podzols** are acidic and have a dark layer of organic matter over a white or grey leached layer. They support coniferous forest vegetation. **Grey-brown Podzolics** are less leached, less acidic, and have less organic matter than Podzols. They support deciduous forest vegetation. **Red-yellow Podzolics** are formed primarily by podzolization and secondarily by laterization. They are brightly colored, well leached, low in organic matter, and high in clay. They support deciduous forest vegetation. **Laterites** are reddish in color, highly leached, very low in organic matter, low in silica and bases, and high in iron and aluminum. They support tropical vegetation. **Prairie** soils have a dark brown or black color, very high organic matter, and high fertility. They support grassland vegetation of the Tall Grass type. The Prairie soil is the last of the pedalfer soils in this listing; the remaining Great Soils are pedocals.

Chernozems have a dark color, high organic matter, high fertility, and a layer of calcium carbonate at about one-meter depth. They support grassland vegetation of the Mixed Grass type. **Brown** and **Chestnut** soils are lighter and less fertile than Chernozems; they have less organic matter and a shallower calcium carbonate layer (0.3 to 1 meter depth). They support grassland vegetation of the Short Grass type. **Grey** soils have a light color, little organic matter, and a calcium carbonate layer at less than 0.3 meter. They support desert vegetation. (Further details on the Great Soils appear in the appropriate chapters of Part 2.)

TOPOGRAPHIC RELIEF

A third major environmental factor that affects vegetation is topographic relief—the varying elevations of land surfaces. Relatively flat areas like much of the Great Plains are said to have low relief and mountainous areas have great or high relief. The interrelatedness of environmental factors becomes apparent when considering the relief factor, since relief affects vegetation primarily through altering other environmental factors, especially climate and soil.

The low hills characteristic of much of the eastern United States provide an example of the importance of topographic relief. These hills have different microenvironments from the top to the bottom of slopes. The hilltops have the driest conditions; water runoff is high, winds are strong (increasing evapotranspiration), soils are least developed, temperatures are high, and solar radiation at the soil surface is at a maximum because of the open vegetation. The bases of the slopes have the wettest microenvironment; water runoff flows to here, winds are slower, soils are better developed, temperatures are cooler, and solar radiation is at a minimum because of the dense vegetation canopy. These environmental factors result in upslope vegetation that is similar to warmer, drier regions and downslope vegetation that is more characteristic of cooler, wetter areas. Midslope positions tend to be intermediate in microenvironment and vegetation. Traditionally, a region is considered to be characterized by its midslope vegetation.

An important factor in areas of varied topographic relief is **slope exposure** (slope aspect), the compass direction a slope faces. In the northern hemisphere, because of the southernly angle of the sun, south-facing slopes receive more solar radiation than north slopes. Consequently, the slopes have different microclimates; the south slopes have higher average temperatures, greater evapotranspiration, and longer growing seasons. On tall mountains, this results in vegetation bands at higher elevations on south-facing rather than on north-facing slopes. On smaller hills, it produces south-slope vegetation related to warmer, drier regions and north-slope vegetation of cooler, wetter areas (Figure 4.8). Species and plant communities may be restricted to south slopes at the northern edge of their range and north slopes at the southern edge.

FIGURE 4.8 The north-south slope effect on a hill in northcentral Colorado in December. The left side of the hill has a north exposure and supports coniferous forest vegetation. The south slope is dominated by grassland vegetation, which is adapted to the warmer, drier microenvironment.

FIGURE 4.9 Map of the physiographic regions of North America.

East–west slope differences in microenvironment and vegetation also occur. These are not based on the amount of solar radiation but on the time of day that direct radiation is received.

Different regions of North America may be described according to their dominant patterns of topographic relief. An understanding of regional vegetation requires a knowledge of these patterns. The major physiographic regions of North America are mapped in Figure 4.9.

BIOTA

The biotic part of the environment of vegetation is another important factor. It includes any organism that affects vegetation, but this statement does not define the biotic factor since all organisms have some effect. However, the role of biotic interactions may be clarified by examples where there are important effects on vegetation. Some examples include the competition between white fir trees and giant sequoias in California's sequoia groves, the invasion of the nonnative (alien) weed halogeton into western rangelands, the buffalo's influence on Great Plains grasslands, the gypsy moth's defoliation of deciduous forests, and the vast impact of the human species.

All biotic influences on vegetation are eventually reducible to the population level. There are several general types of population interactions; three of the more important ones are the following:

1. The two populations are mutually beneficial.
2. The two populations compete.
3. One population serves as a resource for the other.

An example of mutually beneficial populations is the case of plants that have root nodules containing nitrogen-fixing bacteria (page 23). The plants have a supply of nitrate—that which is produced by the bacteria. The bacteria have a food source—the photosynthetic products of the plants. **Mycorrhizae,** associations between fungi and root systems, are a second example of mutualism. The fungi decompose soil organic compounds, making nutrients more available for uptake by plant roots. Plant metabolites are, in turn, a food source for the fungi.

Competitive interaction occurs when coexisting populations require some of the same resources. Competing populations may show a direct or an indirect type of interference. Direct interference involves direct inhibition of each population by the other. **Allelopathy,** the chemical inhibition of plants by each other, is an example of direct interference. Allelopathic plants release chemicals that inhibit the growth and development of others. The indirect type of interference is more common. It occurs when a resource used by both populations is in low supply (relative to needs), as when two plant species are growing on soil that is low in an essential element such as phosphorous.

A third type of population interaction, whereby one population serves as a resource for another, includes all of the many cases of parasitism, predation, and herbivory. For example, buffaloes are herbivores; prior to near extinction in the middle of the nineteenth century, they grazed the taller, eastern grasslands so that they resembled the western Short Grass prairies. The human use of plant populations as resources is an example of this type of population interaction that has had the greatest effect on vegetation.

FIRE

The last major environmental factor to be considered is fire. Its role in influencing vegetation has only been fully acknowledged in the last decade. Through most of the twentieth century, fire was considered an unnatural factor, largely caused by humans and detrimental to ecosystems; however, vegetation scientists, anthropologists, geographers, and others have shown that aboriginal Indians made widespread use of fire as a land-management tool. Most forest, shrubland, and grassland areas of North America were periodically burned to aid in hunting, travel, and the production of wild food crops. Also, lightning was and still is a cause of frequent fires in many areas.

Today, it is recognized that the vegetation of North America evolved with fire as a major environmental factor. One piece of evidence for this is that some species have adaptations such as fire-resistant bark, fire-dependent cone opening mechanisms, and fire-stimulated seed germination. Other evidence is provided by written observations of presettlement fires and by postsettlement vegetation changes in regions where fires have been eliminated.

The effects of fires are highly variable and depend on such factors as the type of vegetation, the season of the year, the time since the last fire, and such weather factors as wind and humidity. All of these factors and more are taken into account when "prescribed" (controlled) fires are used in land management (Figure 4.10). General effects of fires include changes in environmental factors, such as microclimate and soil. If fires disturb the vegetation canopy, the microclimate beneath it is greatly altered; sunlight and temperature fluctuations are increased and humidity and soil moisture are decreased. The direct effect of fires on soils is usually limited to the combustion of organic matter on the soil surface. This alters the availability of many nutrients. Indirect effects can include increased soil erosion.

Fires also obviously affect the biota. The age structures of plant popula-

FIGURE 4.10   A prescribed burn in a restored stand of Tall Grass prairie in western Ohio. Note the tree seedling in the right foreground. Repeated fires result in the exclusion of such woody species and in the dominance of grassland species that are adapted to fire.

tions are modified, often resulting in stands of individuals of the same age. Fires influence the composition of stands; species may be lost (or maintained) and repeated burning leads to dominance by fire-adapted species. Fires also affect the susceptibility of vegetation to insects, diseases, and additional fires. Reoccurring fires result in less intense burning, since combustible fuel never accumulates to high levels.

5

DEVELOPMENT, CLASSIFICATION, AND DISTRIBUTION

DEVELOPMENT

When people think of natural vegetation, they usually picture undisturbed stands, whether it be a massive forest dominated by douglas fir or an open desert with saguaro cactus. However, even prior to European settlement of North America, there were frequent periodic disturbances by animals, storms, fires, and Indians in all types of vegetation. The process by which vegetation recovers following a disturbance or initially develops on an unvegetated site is known as plant **succession;** it is the object of study in the development approach. The different stages of succession are known as **seral stages,** because each is one of the series that, without further disturbance, leads to a **mature stage** (also known as the climax), a stage which is potentially self-perpetuating.

Succession may be difficult to recognize in such formations as the desert and tundra, but it is readily apparent in forest formations where it involves several seral stages, some of which are dominated by different growth-forms (herb, shrub, and tree). An example of succession that is very common throughout most of eastern North America is known as "old-field" succession, since it occurs on agricultural fields that have been abandoned (Figure 5.1). Of the many seeds carried to a field by wind and animal dispersal, only those that are capable of germination and growth on the bare soil surface form the initial community. The first species of old-field succession or any other kind of succession are known as **pioneer species.** The majority of them

FIGURE 5.1 Photographs of stages of old-field succession. (*a*) A two-year-old abandoned field dominated by herbaceous perennial weedy species such as goldenrods. (*b*) An older field with red cedar (the dark evergreen) and osage orange (leafless in early spring). (*c*) An approximately 75-year-old secondary forest dominated by sugar maple, oaks, and other hardwoods. These stands occur less than 10 km apart on the same soil type in southwest Ohio.

FIGURE 5.1   *(Continued)*

are considered "weeds," typical of disturbed habitats. As succession continues, species composition changes and new species become dominant. Nearly all the common species of the early seral stages are herbaceous, but in time various woody species appear and eventually dominate. The woody species also change until, perhaps in two centuries, the relatively stable, self-perpetuating mature stage is reached.

The process by which succession continues from the pioneer species to the mature stage is not fully understood. The traditional view of the succession process holds that as each seral community affects a site through shading, litter fall, chemical exudates, and so on , the microenvironment is modified and species that are better adapted to the new conditions invade and competitively replace those already present. For example, pioneer species begin growth on bare, warm, dry sites. However, they add organic cover, increase shade, raise soil-moisture levels, and are succeeded by other species better adapted to this altered microenvironment. Additional species replacements continue until the mature stage is reached. It is self-perpetuating, because its species are capable of reproducing beneath their own canopy.

Today, it is generally recognized that the causes of succession involve additional, possibly more important, factors. For example, there is evidence that succession does not always depend on the continued invasion of new species. At times, species characteristic of all seral stages may be present very early in succession. In these cases, observed changes in dominants are the result of different rates of growth. Herbs, which reach maximum size in one or two years, would be expected to dominate the early stages of succession, followed by slower maturing shrubs and later by trees.

Various factors influence the species composition of different stages and thereby affect the direction of succession. These factors include climate, soil, topographic relief, and so forth, as discussed in Chapter 4. Chance also may play a major role. For example, the dispersal of seeds is often a chance process, and because of a single storm, certain species may be dispersed into a site of succession, while others are not. The direction of succession is also greatly influenced by which plants survive a disturbance and are therefore present at the start of succession. For example, the species that survive a forest fire either as mature plants or as seeds in the soil may become important components of at least the early seral stages.

These problems in understanding succession are compounded by the fact that successions are influenced not only by autogenic (self-induced) factors such as shading, but also by allogenic (external) factors such as climatic change. It is often virtually impossible to separate the effects of these two categories of factors. Examples of the effects of long-term climatic shifts on the vegetation of North America are described in the next chapter.

Up to this point, only secondary successions have been discussed. **Secondary successions** are those that occur following a disturbance of the vegetation that leaves some organisms and soil in place. **Primary successions** occur on totally unvegetated areas that lack a soil; examples include bare-rock outcrops in mountains, sand dunes along coastlines, and gravel-outwash plains of glaciers.

Successions may also be classified according to the moisture content of the substrate. Succession on rock outcrops is an example of **xerarch succession**—that occurring on a dry substrate. **Hydrarch succession** occurs on a wet substrate; an example is the gradual filling in of a pond with aquatic plants, followed by the formation of a bog and the further succession of terrestrial plants. The third type of succession, **mesarch succession,** occurs on soils of intermediate soil-moisture levels; an example is old-field succession.

All types of succession are characterized by continuous vegetation change (up to the mature stage); nevertheless, most vegetation scientists describe the varying plant communities as individual seral stages. Each stage is usually named for the dominant plants. An example of primary succession that has been studied for most of this century occurs on recently deglaciated areas in Alaska (Figure 5.2). Three stages have been described: a moss-perennial herb-prostrate willow stage, a willow-alder thicket stage, and a spruce-hemlock forest stage. The stages may be referred to as pioneer, young, and mature phases of succession. All successions can be divided into these phases. Those with more than three stages have one that corresponds to the pioneer phase, several in the young phase, and one as the mature phase.

Of course, there are major differences between the pioneer, young, and mature phases. All seral stages differ from the mature or climax stage in that, by definition, they are characterized by directional change, but the mature stage is relatively stable. The occurrence of directional change cannot be taken to imply that the pathway of succession is a rigid one. All stages include numerous, confusing, nondirectional changes. For example, variations in annual precipitation can cause fluctuations in the vegetation of any stage, and prolonged drought could reverse vegetation change from that which is expected. Nevertheless, barring a disturbance, succession progresses directionally from the pioneer through the young to the mature phase. Directional change would not be observable within a mature stage unless there was directional change in the environment (i.e., in allogenic factors such as climate).

In some instances, the mature stage may show cycles of change. A common observation in many eastern forests is the microsuccession occurring beneath gaps formed in the canopy when a large tree falls. A small gap may be filled by lateral growth of adjacent mature trees; however, large gaps allow the growth of saplings into the canopy. Thus, some species may be maintained in the canopy by the periodic forming and filling of large gaps. In these situations, the composition of the mature stage is dependent on small-scale cycles of change.

In addition to the difference of directional change in seral stages and nondirectional change in the mature stage, there are various other differences between the stages (see E. P. Odum, Suggested Readings, following Chapter 6). Many of these are actually hypotheses and need further validation; nevertheless, they provide insight into succession. From a composition standpoint, (total) species diversity is thought to increase throughout succession, peaking in a late seral stage (in which there is a mixture of seral and climax species). However, plant species diversity frequently appears to decrease in succession, resulting in a mature community dominated by only a few species. Vegetation structure usually increases in height (as individuals are of greater size), and stratification becomes more apparent (compare Figure 2.5, top, and Figure 2.6).

Functionally, food webs are thought to become more complex during succession, and there is greater energy flow through detritus food chains. In addition, both gross primary production and total community respiration increase, until in the mature stage they are thought to be approximately equal. Thus, net community production, which is believed to peak in a young seral stage, would be zero in the mature stage. The ratio of gross primary production to standing crop biomass apparently decreases as an

FIGURE 5.2 Photographs of stages of primary succession on deglaciated areas in Glacier Bay, Alaska. (*a*) The moss-perennial herb-prostrate willow pioneer community surrounding a stand of the taller, more mature willow-alder thicket community. (*b*) Tree invasion in a thicket community. (*c*) The mature forest community dominated by Sitka spruce, western hemlock, and mountain hemlock. *Source*. Cooper, W. S. 1939. A fourth expedition to Glacier Bay, Alaska. Ecology 20:130–155.

increasing proportion of the biomass of the vegetation is made up of supporting nonphotosynthetic tissues, such as those of wood and bark. The total amount of organic matter (living and dead) increases throughout succession. It is thought that decomposers become more important in providing minerals for plant growth in the later stages. As a result, mineral cycles are thought to become tighter; that is, less is lost from the biological component

FIGURE 5.2 *(Continued)*

and nutrient conservation is high. Lastly, it is hypothesized that interactions between populations become more highly developed and complex throughout succession.

CLASSIFICATION

In the classification of vegetation, plant ecologists seek to categorize units of vegetation, that is, to name them and to group them according to similarities. An example already discussed is the classification of stands as pioneer, young, or mature based on their successional status. There are several other criteria that may be used in classifying vegetation. In fact, one could implement any of the approaches to the study of vegetation as classification methods.

Methods based on composition rely on species presence, usually with some measure of abundance such as density and/or one of the size factors (e.g., cover, basal area, etc.). Methods involving vegetation structure may emphasize differences in stratification. One method of classification based on function contrasts biogeochemical cycling in different ecosystems. The relationships between vegetation classification and such environmental factors as climate and soil have already been mentioned. Most classifications consider development, looking at the successional age of the vegetation. Classification that involves the distribution of vegetation is an obvious approach. Those systems based on history emphasize the common origin of stands.

Thus, different methods of classification emphasize different approaches to the study of vegetation. Most classification systems do not employ a single approach, but instead combine two or more.

There is no universal method of classifying vegetation. The question is controversial. Not only do vegetation scientists disagree on methodology, but some believe that the near infinite variation in vegetation makes it impossible to construct meaningful classifications beyond general categories. Regardless of the level of detail, however, classifications are based on relative, not absolute, discontinuities in vegetation.

The system of classification used in this book stresses composition and structure and considers successional development. Mature vegetation is emphasized, but seral stages are mentioned if they were common in presettlement times. The hierarchy of classification is stand–association–formation. As discussed in Chapter 1, stands are actual examples of relatively homogeneous vegetation; associations are groups of compositionally and structurally similar stands; and formations are groups of associations dominated by the same growth-form.

DISTRIBUTION

Since the distribution of vegetation cannot be described without first defining categories or types, classification is an obvious prerequisite to the study of distribution. The major method of describing vegetation distribution is through maps (Figure 5.3). Vegetation maps are available for most of the world, although many regions have not been mapped in detail. European countries, in particular, have done extensive vegetation mapping. In North America, only the state of California has had a program of detailed mapping. Since vegetation is an indicator of ecological conditions, these maps aid land-use planning.

Several precautions must be stated about vegetation maps. First, different maps use different systems for classifying vegetation. Second, some maps represent presettlement vegetation, others current vegetation, and still others potential vegetation (that which presumably would develop if human influence ended). Third, the boundary lines drawn between vegetation types usually cannot be interpreted as representing abrupt vegetation transitions (ecotones) that are observable in the field. Vegetation transitions in undisturbed areas are usually gradual, paralleling the gradual changes of various environmental factors. Abrupt vegetation transitions usually can be correlated with abrupt changes (steep gradients) in one or more environmental factors. Lastly, even if an area is shown as a single vegetation type, it cannot always be considered homogeneous. There is great vegetational variation within any area. Only relative continuities (and discontinuities) can be

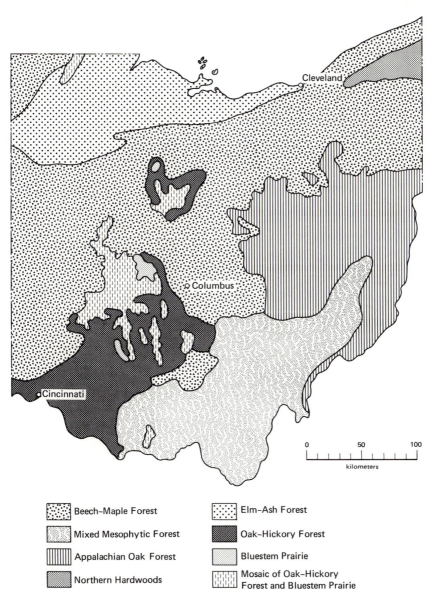

Legend:

Beech–Maple Forest   Elm–Ash Forest

Mixed Mesophytic Forest   Oak–Hickory Forest

Appalachian Oak Forest   Bluestem Prairie

Northern Hardwoods   Mosaic of Oak–Hickory
Forest and Bluestem Prairie

FIGURE 5.3   A vegetation map of Ohio. *Source.* Redrawn from Küchler, A. W. 1964. Potential natural vegetation of the conterminous United States. American Geographical Society, Special Publication 36, New York. Map. The vegetation classification is somewhat different from that used in Part 2.

illustrated on maps, which are by nature only broadly representative. The scale of the map and the skills of the vegetation scientist and the cartographer largely determine whether a map accurately depicts vegetation and its distribution.

A general vegetation map of North America appears in the frontispiece. Somewhat more detailed maps of North America exist, but significantly greater detail is available only for specific countries. For example, the potential natural vegetation of the United States (exclusive of Alaska and Hawaii) has been mapped by A. W. Küchler (see the Suggested Readings list that follows Chapter 6). Similar maps are not available for all of North America, but more detailed maps have been made of states, provinces, and smaller areas (see Küchler and McCormick, Suggested Readings).

6

HISTORY

History, the eighth approach to vegetation, studies the role of past events in the origin of vegetation types. Since both the history and the development approaches deal with vegetation change, the two are sometimes confused. History, however, covers extremely long time periods and considers only allogenically induced changes. In contrast, development focuses on the relatively short-term change of succession in which autogenic factors are involved.

METHODS

Plants have inhabited land for more than 500 million years, so there are great problems in reconstructing past vegetation types. The major method for studying vegetation of an extremely old age is by the analysis of plant fossils. These form when parts such as leaves are covered by the sediments of a river, for example, and the shape of the structures is preserved as the surrounding sediment is changed to rock. Fossils may be dated by analysis of the radioactivity of the rock containing them. Information from archeological sites can sometimes be used to supplement that gained from fossils. For example, the presence of bones of grassland animals would indicate the prior existence of grassland vegetation in the region.

Unfortunately, fossil study does not provide a complete record of the history of vegetation. The formation and preservation of fossils is a chance process. Some locations, such as uplands away from sites of sedimentation, are not conducive to fossil formation, so the plants of these areas are poorly represented by fossils.

One method of analyzing the vegetation history of more recent (but unrecorded) time is pollen analysis. Most dominant species are wind polli-nated, and their pollen is continually falling to the ground. In sites where pollen is preserved for long periods of time, such as bogs and lakes, there is a record of past vegetation. A coring device is used to remove a vertical section of the sediments of a bog or lake, and this is analyzed in increments to see how the proportions of pollen grains of different species (or genera) have changed.

Pollen studies have shortcomings; for example, different species produce different amounts of pollen, pollen may be blown in from other areas, pollen of some species may not preserve well, and pollen of several species may be too similar for differentiation. Nevertheless, pollen analyses have provided valuable information on vegetation history, especially in the more mesic regions of North America.

A useful method for studying the vegetation history of the arid regions of North America involves examination of plant materials contained in wood rat middens (deposits). Wood rats accumulate masses of plant materials that are preserved by a covering of dried urine. Study of the leaves, seeds, fruits, and other plant parts that are found in these middens reveals the composi-tion of the surrounding vegetation at the time of collection. The age of wood rat middens can be determined by radiocarbon dating; some are over 40,000 years old. Most wood rat studies have been done in desert regions where arid conditions favor the preservation of middens; however, given the wide range of wood rats (coast to coast and from northwest Canada to Nicaragua), this technique should prove useful in more mesic regions if ancient middens can be found in protected sites such as dry caves.

Some relatively recent vegetation changes may be studied by tree-ring analysis. Trees usually produce a single growth ring each year, the width of which reflects environmental conditions during the growing season. Tree rings may be observed on cut stumps or logs, but a nondestructive method involves the use of an increment borer—a narrow steel tube that is screwed into a tree and when removed encloses a cross section of the trunk. Tree-ring analyses can be used to date individual trees, fires (if fire scars formed in the wood), droughts, and other major environmental changes. Studies in the southwestern United States have produced a tree-ring record for bristlecone pine that dates back over 8200 years. This was accomplished by cross-dating individual living and dead, undecayed trees by matching sequences of wide and narrow rings. Archaeologists have used cross-dating to determine the age of old Indian villages where structures contained wooden supports.

There are several methods of studying postsettlement vegetation history. Sometimes it is possible to use the written observations of explorers and early settlers. However, these descriptions are usually so general that they are valuable primarily where major vegetation changes have occurred. Land surveys at the time of settlement provide more detailed information on some areas of North America. Surveyors usually described the vegetation in order that possible land uses could be predicted from their survey notes. In many areas the surveyors also recorded the species of "witness trees" that marked their survey lines. A picture of past vegetation can be reconstructed from such detailed surveys, provided that the surveyor was capable of accurate species identification and that the survey was not fraudulent.

In forested regions, the records of logging companies also can be used to determine the nature of the original vegetation. Old photographs can also aid in the study of historical vegetation change in some areas. Finally, preserved ("relict") stands are used in studies of presettlement vegetation. Unfortunately, suitable areas in some regions are small and uncommon. Also, many supposed relict areas have in fact greatly changed as a result of postsettlement human influence.

NORTH AMERICAN VEGETATION

The many vegetation types of North America are either dominated by angiosperms (the flowering plants) or have them as a major component. This was not always the case. Angiosperms first became abundant late in the Cretaceous Period, so this discussion of the history of North American vegetation covers changes through the Tertiary Period to the present. Events of this approximately 65-million-year time span can best be understood by interrelating changes in three factors: relief, climate, and vegetation.

At the beginning of the Tertiary Period the North American continent was greatly different from what it is today, not only in shape, but more importantly in topographic relief. In general, the continent was relatively flat. The old age Appalachian and newly developing Rocky Mountains were low in elevation. The Sierra Nevada, Cascade, and Coast Range mountains had not yet begun to form. The climate was also less varied than it is today. It has been characterized as equitable (i.e., warm with ample precipitation) throughout most of the continent. Although our knowledge of the vegetation at the beginning of the Tertiary Period is limited, it appears that with the relatively uniform environment the vegetation was less diverse. Three **geofloras** (major vegetation units whose essential characteristics are maintained through space and time) have been described as the Arcto-Tertiary, the Neotropical-Tertiary, and the Madro-Tertiary (Figure 6.1).

The **Arcto-Tertiary Geoflora** covered the northern half and largest land area of the continent. It was a rich forest of mesic species, a mixture of gymnosperms and deciduous angiosperms. It resembled two types of present-day forests: that found in the coastal region of northern California (the redwood-dominated stands of the Northwest Coastal coniferous forest; see page 124) and that of the low elevations of the southern Appalachian Moun-

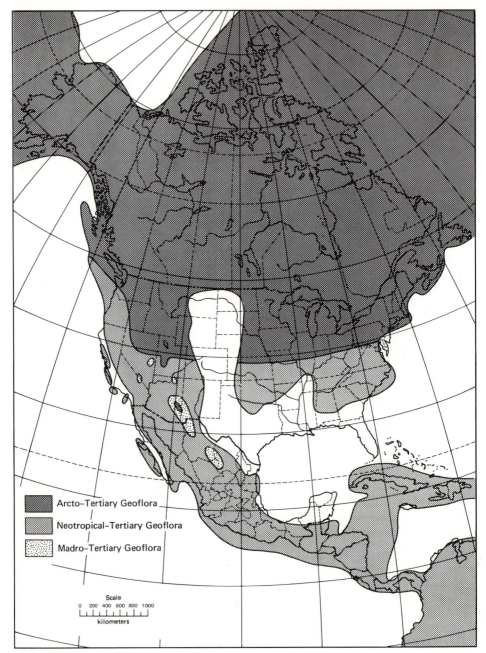

FIGURE 6.1   Distribution of geofloras in the early Tertiary Period (approximately 65 million years ago). The continent's shape is given by the outline of the geofloras. Also, the continent's latitudinal and longitudinal position was different than it is today. *Source*. Redrawn from a map of D. I. Axelrod that appeared in Raven, P. H., R. F. Evert, and H. Curtis. 1976. Biology of plants. 2nd ed. Worth, New York. 685 p.

tains and Plateau (the cove hardwood stands of the Mixed Mesophytic deciduous forest; see page 137). The Arcto-Tertiary Geoflora had three major elements: (1) species that today are characteristic of eastern North America,
including hickories, beech, elms, maples, and ashes; (2) species more common today in western North America such as firs, cedars, spruces, sequoias,
pines, other maples, alders, and so on; and (3) east Asian species such as
ailanthus, ginkgo, dawn redwood, and many others.

The **Neotropical-Tertiary Geoflora** extended over most of the southern
half of North America. It was made up of large rich forests similar to many of
today's subtropical and tropical forests that are found between southern
Mexico and Panama. Most of its species were broad-leaved evergreen angiosperms adapted to consistent, high rainfall and mild temperatures.

The third geoflora, the **Madro-Tertiary,** covered a few, relatively small
areas in the southwestern United States and adjacent Mexico. This geoflora
was of more recent origin, being derived from the other two. It occurred only
in relatively dry areas (which at that time were not extensive). It consisted of
a variety of vegetation types, including woodland, shrubland, and desert
grassland. The dominant growth-forms were shrubs and trees with somewhat small, strong, drought deciduous leaves. Many of the relict species of
this geoflora presently occur in the summer rain areas of central Arizona and
northern Mexico (including the Sierra Madre mountains for which the geoflora is named).

In summary, North America at the beginning of the Tertiary Period was a
relatively flat continent with a mild climate. It supported three general types
of vegetation, only two of which had a wide distribution.

North American environments became much more varied between the
early and middle Tertiary. Some change occurred in the Appalachian Mountain region where a series of small uplift and erosion cycles formed more
diverse topography. The major change in relief was the continued rise of the
Rocky Mountains. The climate throughout the continent apparently became
cooler and drier. Greater topographic relief produced low temperatures at the
high elevations and a rain shadow east of the Rocky Mountains.

These changes in the environment produced changes in the vegetation.
The Arcto-Tertiary Geoflora extended southward as the climate cooled.
Regional vegetational variations developed, such as in western North America where cooler temperatures led to the loss of many deciduous angiosperm
species. The Neotropical-Tertiary Geoflora, being adapted to very mild conditions, migrated southward and coastward. The area it covered greatly
decreased because of less land mass in southern North America. The Madro-
Tertiary Geoflora, however, expanded in area as arid conditions became
more widespread. Finally, a new type of vegetation, the grasslands, developed in the relatively dry rain shadow region of the uplifting Rocky Mountains. It did not resemble any of the three geofloras in appearance.

In the late Tertiary Period, the Appalachian and Rocky Mountains reached their present height. In addition, the uplift of the Sierra Nevada, Cascades, Coast Ranges, and Great Basin began and proceeded at a geologically rapid pace. These events led to further climatic variation. The entire continent became still cooler and drier. The high elevations and latitudes developed cold climates with short growing seasons. The two tall western mountain chains produced rain shadows to their east. Lastly, the proportion of precipitation received during the summer decreased west of the Rocky Mountains, especially near the Pacific Coast, where a summer dry, winter wet (Mediterranean) climate developed.

The vegetation responded to these changes. Differences between the east and west portions of the Arcto-Tertiary Geoflora became more pronounced. In the more mesic east, forests were composed primarily of deciduous angiosperms. Western forests became increasingly dominated by gymnosperms and were restricted to mesic sites, such as those along the coastal fog belt and at the medium and high elevations of mountains. The Madro-Tertiary Geoflora also diversified. Many of its species became concentrated in central Arizona and portions of Mexico that had summer rain. Other of its species became adapted to the summer dry conditions of the California region where they formed chaparral shrubland (page 213) and related vegetation. The products of the Madro-Tertiary Geoflora are dominated by shrubs and/or small trees with the broad-sclerophyll growth-form (flat, hard leaves). The Neotropical-Tertiary Geoflora continued its southward, coastward migration with the further cooling of the climate. It produced various types of tropical vegetation.

In addition to the products of the geofloras, grassland vegetation expanded east of the Rocky Mountains and in a few other areas. Tundra vegetation appeared on the highest mountain summits and in the far north, areas of extreme winter cold and cool, short growing seasons. Lastly, a desert formation developed east of the Sierra Nevada and in eastern Mexico, locations with very low annual precipitation. The desert is related primarily to the Madro-Tertiary Geoflora and secondarily to the Arcto-Tertiary Geoflora.

Following the end of the Tertiary (two to three million years ago), the early Quaternary Period witnessed the development of topographic relief, climates, and vegetation very similar to the present. In more recent, but prehistorical, times, two new factors greatly affected vegetation. One of these was the migration of the human species into North America. As the human population grew and as hunting and agricultural practices improved, human impact on vegetation increased. Especially important was the aboriginal Indians' widespread use of fire.

The second important, relatively recent factor was continental glaciation. At four different times during the Pleistocene portion of the Quaternary, climatic changes resulted in long-term accumulation of snow in the north.

Great masses of snow and ice built up, and their enormous weight caused the slow southernly flow of continental ice sheets of over a thousand meters in thickness. The rate of movement was very slow; each of the four Ice Ages lasted several thousands of years. The glaciers physically altered landscapes, destroying vegetation, flattening topographic relief, and changing river drainages. Climates were also modified by the advancing ice sheets.

The continental and mountain glaciers had great effects on vegetation. The northernmost geoflora, the Arcto-Tertiary, was especially affected. Some Arcto-Tertiary species and genera became extinct in North America; examples include the dawn redwood and ginkgo, species which had once been widespread (today they have been reintroduced into North America as ornamentals). Other species were divided into two isolated populations, one in the deciduous forest region of the eastern United States and the other in the coniferous forest region of the West. With adaptation to local conditions, the populations evolved into separate species. Today, the two regions, for example, have different but similar species of dogwoods, redbuds, and sycamores. Such speciation did not greatly alter the general appearance of the vegetation of the two regions, but the differences and similarities of the two floras are of great botanical interest.

The most recent of the four Ice Ages, the Wisconsin, peaked approximately 18,000 years ago. Its maximum advance in the eastern United States is shown in Figure 6.2, along with the distribution of vegetation south of its

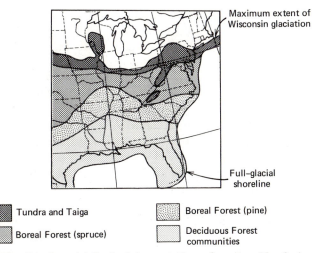

FIGURE 6.2 Distribution of full-glacial vegetation of eastern North America (tentative reconstruction). Despite the banding shown on the map, zonal migration of vegetation did not occur. Also, evidence indicates that vegetation units of glacial age were different from analagous vegetation types that occur today. *Source.* Redrawn from Whitehead, D. R. 1973. Late-Wisconsin vegetational changes in unglaciated eastern North America. Quaternary Research 3:621–631.

boundary. Reconstruction of the full-glacial vegetation is based on pollen analyses from sites south of the glacier. This Ice Age ended 10,500 years ago.

Pollen studies done in glaciated areas give a postglacial vegetation history of the region. Studies in Ohio have revealed several zones of pollen. Although tundralike communities may have had a local distribution in the area (and were more widespread elsewhere in the eastern United States), the lowest (oldest) pollen zone typical of the Ohio region indicates a period of spruce-fir vegetation. Today, similar vegetation occurs hundreds of kilometers north, in the Boreal coniferous forest of Alaska and Canada. Above the spruce-fir pollen zone is one dominated by pine pollen. This indicates a warming and drying of the climate. The zone of pine pollen is typically followed by one of the deciduous forest species; here beech was dominant.

The next higher pollen zone was produced during the Hypsithermal, a time of warm, dry conditions that is thought to have been at a maximum in Ohio approximately 3000 years ago. Pollen from this time indicates a reduction in beech, a species adapted to mesic conditions, and increases in oaks and hickories, genera which include many species adapted to dry conditions. With the end of the Hypsithermal, the cooler, moister climate apparently led to the development of a deciduous forest with the composition of presettlement stands. In Ohio, beech pollen is again common in this zone (much of Ohio is in the Beech-Maple association of the deciduous forest; see page 141). Lastly, there is a sudden increase in the pollen of weedy species in the upper layers. This marks the initiation of forest clearing and agriculture that accompanied pioneer settlement.

Southern displacement of species also occurred in the central and western portions of the continent. At maximum Wisconsin stage glaciation, stands of Boreal coniferous forest species occupied the northern Great Plains. There is evidence that the southern portion of this physiographic region supported stands of pines. South of the Great Plains in what is now the Chihuahuan Desert of Mexico, there was woodland vegetation of pinyon pine, evergreen oaks, and juniper. Wide expanses of open grassland vegetation are not known to have been present during maximum glaciation. Instead, postglacial fires in the Great Plains produced open grassland vegetation on areas of flat or rolling terrain, leaving forest and woodland trees restricted to riparian habitats and topographic breaks such as escarpments.

Further west, greater topographic diversity resulted in more complex vegetation changes. For example, coniferous forests were dominant in the lower elevations of the Great Basin physiographic region at the peak of glacial activity. With postglacial climatic change, these forests became restricted to their present locations at high elevations in isolated mountain ranges. The forest was replaced by coniferous woodland and desert vegetation at lower elevations. The woodland species migrated from the Mojave Desert region, an area of lower elevation to the south. The (Great Basin) desert species migrated from glacial refugia whose locations are unknown.

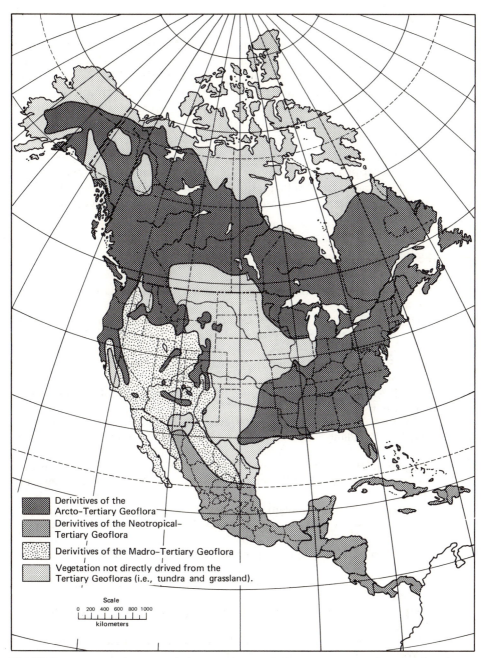

Derivitives of the
Arcto–Tertiary Geoflora

Derivitives of the Neotropical–
Tertiary Geoflora

Derivitives of the Madro–Tertiary Geoflora

Vegetation not directly drived from the
Tertiary Geofloras (i.e., tundra and grassland).

Scale

0   200   400   600   800 1000

kilometers

FIGURE 6.3 Distribution of the products of the Tertiary geofloras. The desert formation has been included with the Madro-Tertiary Geoflora, but it also has some floristic relationships with the Arcto-Tertiary Geoflora. Compare this map with the frontispiece and with Figure 6.1.

In summary, the history of the formations of North America extends back to the early Tertiary Period. Changes in the vegetation parallel changes in the climate, which reflect the changes in topographic relief. The Arcto-Tertiary Geoflora produced the deciduous and coniferous forest formations (Figure 6.3). The Madro-Tertiary Geoflora developed into various types of broad-sclerophyll dominated vegetation in the southwestern United States and adjacent Mexico. The Neotropical-Tertiary Geoflora diversified into several types of tropical vegetation. Also developing since that time but not greatly resembling the appearance of any of the three Tertiary gefloras, were the grassland, desert, and tundra formations. Glaciation and activities of aboriginal Indians were important, somewhat recent influences on vegetation.

SUGGESTED READINGS FOR FURTHER STUDY

Ashby, M. 1969. An introduction to plant ecology. 2nd ed. St. Martin's, New York. 287 p.

Axelrod, D. I. 1977. Outline history of California vegetation, p. 139–193. *In* M. G. Barbour and J. Major (eds.). Terrestrial vegetation of California. Wiley, New York.

Bailey, R. G. 1976. Ecoregions of the United States. USDA Forest Service Intermountain Region, Ogden, Utah. Map.

———. 1978. Description of the ecoregions of the United States. USDA Forest Service Intermountain Region, Ogden, Utah. 77 p.

Basile, R. M. 1971. A geography of soils. Brown, Dubuque, Iowa. 152 p.

Billings, W. D. 1978. Plants and the ecosystem. 3rd ed. Wadsworth, Belmont, California. 177 p.

Braun-Blanquet, J. 1932. Plant sociology: The study of plant communities. Translated, revised, and edited by G. D. Fuller and H. S. Conrad. McGraw-Hill, New York. 439 p.

Cain, S. A. 1944. Foundations of plant geography. Harper & Row, New York. 556 p.

Cain, S. A. and G. M. de O. Castro. 1959. Manual of vegetation analysis. Harper & Row, New York. 325 p.

Dansereau, P. 1957. Biogeography: An ecological perspective. Ronald, New York. 394 p.

Daubenmire, R. 1968. Plant communities: A textbook of plant synecology. Harper & Row, New York. 300 p.

———. 1974. Plants and environment: A textbook of autecology. 3rd ed. Wiley, New York. 422 p.

———. 1978. Plant geography: With special reference to North America. Academic, New York. 338 p.

Drury, W. H. and I. C. T. Nisbet. 1973. Succession. Journal of the Arnold Arboretum 54:331–368.

Egler, F. E. 1977. The nature of vegetation, its management and mismanagement: An introduction to vegetation science. Aton Forest, Norfolk, Connecticut. 527 p.

Eyre, S. R. 1968. Vegetation and soils: A world picture. 2nd ed. Aldine, Chicago. 328 p.

Gates, D. M. 1972. Man and his environment: Climate. Harper & Row, New York. 175 p.

Gleason, H. A. and A. Cronquist. 1964. The natural geography of plants. Columbia University Press, New York. 420 p.

Good, R. 1974. The geography of the flowering plants. 4th ed. Longman, London. 557 p.

Horn, H. S. 1975. Forest succession. Scientific American 232(5): 90–98.

Kellman, M. C. 1975. Plant geography. St. Martin's, New York. 135 p.

Kershaw, K. A. 1974. Quantitative and dynamic ecology. 2nd ed. American Elsevier, New York. 308 p.

Kormondy, E. J. 1976. Concepts of ecology. 2nd ed. Prentice-Hall, Englewood Cliffs, New Jersey. 238 p.

Kozlowski, T. T. and C. E. Ahlgren (eds.). 1974. Fire and ecosystems. Academic Press, New York. 542 p.

Küchler, A. W. 1964. Manual to accompany the map. American Geographical Society, Special Publication 36, New York. 116 p.

———. 1964. Potential natural vegetation of the conterminous United States. American Geographical Society, Special Publication 36, New York. Map.

———. 1967. Vegetation mapping. Ronald, New York. 472 p.

Küchler, A. W. and J. McCormick. 1971. Vegetation maps of North America. Van Bekhoven, Naarden, Netherlands. 453 p. Reprint of the 1965 edition published by The University of Kansas Libraries, Lawrence.

Lieth, H. 1975. Primary production of the major vegetation units of the world, p. 203–215. *In* H. Lieth and R. H. Whittaker (eds.). Primary productivity of the biosphere. Springer-Verlag, New York.

Lieth, H. and R. H. Whittaker (eds.). 1975. Primary productivity of the biosphere. Springer-Verlag, New York. 339 p.

Likens, G. E., F. H. Bormann, R. S. Pierce, J. S. Eaton, and N. M. Johnson. 1977. Biogeochemistry of a forested ecosystem. Springer-Verlag, New York. 146 p.

Mueller-Dombois, D. and H. Ellenberg. 1974. Aims and methods of vegetation ecology. Wiley, New York. 547 p.

Odum, E. P. 1971. Fundamentals of ecology. 3rd ed. Saunders, Philadelphia. 574 p.

Oosting, H. J. 1956. The study of plant communities: An introduction to plant ecology. 2nd ed. Freeman, San Francisco. 440 p.

Phillips, E. A. 1959. Methods of vegetation study. Holt, Rinehart and Winston, New York. 107 p.

Polunin, N. 1960. Introduction to plant geography and some related sciences. McGraw-Hill, New York. 640 p.

Ricklefs, R. E. 1979. Ecology. 2nd ed. Chiron, New York. 966 p.

Schimper, A. F. W. 1903. Plant-geography upon a physiological basis. Translated by W. R. Fisher. Revised and edited by P. Groom and W. R. Fisher. Clarendon, Oxford. 839 p.

Shimwell, D. W. 1972. The description and classification of vegetation. University of Washington Press, Seattle. 322 p.

Smith, R. L. 1974. Ecology and field biology. 2nd ed. Harper & Row, New York. 850 p.

Stålfelt, M. G. 1972. Stålfelt's plant ecology: Plants, the soil and man. Translated by M. S. Jarvis and P. G. Jarvis. Wiley, New York. 592 p.

Steila, D. 1976. The geography of soils. Prentice-Hall, Englewood Cliffs, New Jersey. 222 p.

Tall Timbers Fire Ecology Conference. 1962– . Proceedings. Tall Timbers Research Station, Tallahassee, Florida.

Thomas, W. L. (ed.). 1956. Man's role in changing the face of the earth. University of Chicago Press, Chicago. 1193 p.

Walter, H. 1973. Vegetation of the earth in relation to climate and eco-physiological conditions. Translated by J. Wieser. Springer-Verlag, New York. 237 p.

Walter, H., E. Harnickell, and D. Mueller-Dombois. 1975. Klimadiagramm-karten. Fisher, Stuttgart, West Germany. 36 p. + maps.

Walter, H. and H. Lieth, 1967. Klimadiagramm-weltatlas. Fisher, Jena, East Germany.

Warming, E. 1909. Oecology of plants; an introduction to the study of plant communities. Assisted by M. Vahl. Prepared for publication in English by P. Groom and I. B. Balfour. Oxford University Press, Oxford. 422 p.

Weaver, J. E. and F. E. Clements. 1938. Plant ecology. 2nd ed. McGraw-Hill, New York. 601 p.

Wells, P. V. 1970. Postglacial vegetational history of the Great Plains. Science 167:1574–1582.

———. 1976. Macrofossil analysis of wood rat *(Neotoma)* middens as a key to the vegetational history of arid America. Quaternary Research 6: 223–248.

Whittaker, R. H. 1975. Communities and ecosystems. 2nd ed. MacMillan, New York. 385 P.

Wiens, J. A. (ed.). 1972. Ecosystem structure and function. Proceedings, Thirty-first Annual Biology Colloquium. Oregon State University Press, Corvallis. 176 p.

PART 2

THE MAJOR FORMATIONS OF NORTH AMERICA

The vegetation of North America is discussed in the following chapters. A map of the major formations appears in the frontispiece. Each chapter begins with a series of photographs that introduce the vegetation. This is followed by information such as details of distribution, including a map, and a synopsis of the history of the formation.

The environment of the regions covered by the formation is discussed in depth. Reference to Chapter 4—especially the maps of temperature, precipitation, soils, and physiographic regions—will prove helpful. Climate diagrams illustrate the range of conditions within a formation.

Each chapter includes a general description of the vertical stratification of the vegetation. Composition is dealt with in terms of the major dominants. Scientific names are included the first time a species is mentioned in the text of a chapter; only common names are used thereafter. A complete list of common names with their scientific equivalents appears in the Appendix.

Although emphasis is on the formations, their major subdivisions, and the species that dominate them, it must be recognized that spatial variations in the environment result in variations in vegetational composition and structure, even within subdivisions.

Additional insight into the relationship between vegetation and environment is provided by the sections on species adaptations that appear in each chapter.

A concluding part of the chapters describes human impact on the formations. These sections illustrate human dependency on vegetation and furnish information about major threats to natural systems.

Each chapter is followed by a list of recent review articles and important research papers. These contain detailed information on various topics, including subdominant species, vegetation variations, and species adaptations, as well as references to numerous other important studies.

The format of Chapter 13 differs slightly from that of the other chapters; a regional approach is used and several diverse formations are included under the heading of Tropical Vegetation. These formations are covered as subdivisions of the chapter and include the rain forests, montane and seasonal vegetation types, and tropical savannas.

7

TUNDRA VEGETATION

North American tundra vegetation occurs across the northernmost portion of the continent and extends southward along mountains as far as Guatemala. The tundra in the north is Arctic tundra, and that of the mountains is Alpine tundra. Arctic tundra is divided at about 72–73° north latitude into the Low Arctic to the south and the High Arctic to the north.

The most striking aspect of tundra vegetation is the lack of trees (treeline is discussed in Chapter 8, page 106). Without the tree growth-form, the structure of the tundra is dominated by herbs or shrubs. In areas of a moderate environment the herb dominants are species of grasses and sedges (*Carex* spp.); also present are various mosses, lichens, and forbs (broad-leaved herbs). Where soil moisture is not limiting, the herbaceous cover is well developed and, in very wet sites, grass-sedge bogs and meadows are common. In drier or disturbed areas, vegetation cover is less continuous, and forbs tend to predominate. Many sites, especially in alpine regions, are extremely rocky, and plants grow mainly between the rocks and boulders. These areas are known as fell-fields. Many lichens and some mosses grow on the rocks. Lichens are found throughout the tundra, but are dominant in the most extreme habitats, such as on exposed ridges. In contrast, shrubs are most common in the mildest environments. Dominants include willows (*Salix* spp.) and heaths, such as blueberries (*Vaccinium* spp.). Shrubs also may be found in more extreme environments, but there they are dwarfed to a few centimeters and are often restricted to sites not exposed to strong winds.

Usually, the shrub, herb, and surface (moss-lichen) layers are not found together. Instead, except in arid areas, there is much community diversity with varying growth-form dominance, that is, shrub communities may be intermixed with herbaceous communities and moss-lichen dominated communities. There is little evidence that these are successionally related.

(a) Alpine meadow and lake in Sequoia National Park, California.
(b) Alpine tundra and timberline in Sequoia National Park, California.

(c) Aerial view of Arctic tundra landscape on Richards Island, Northwest Territories. *Source*. L. C. Bliss, unpublished.

(d) Arctic tundra in the Caribou Hills, near Inuvik, Northwest Territories. *Source*. L. C. Bliss, unpublished.

Apparently, the tundra environment is such that the vegetation has less impact on it than, for example, a forest has on altering its environment. Nevertheless, the microenvironments produced by tundra plants are extremely important for their growth and reproduction.

DISTRIBUTION

The distribution of Arctic tundra vegetation is illustrated in Figure 7.1. The influence of climate is shown by the southward extension of the tundra in the east where a continental climate is well developed. Ocean currents also have an effect; the Japan Current warms the west coast, and the Labrador Current cools the east coast. The Alpine tundra covers little land area, so its distribution cannot be shown on a map of this scale. It is restricted to higher elevations at lower latitudes. Alpine tundra is found along the Cascade and Sierra Nevada Mountains as far south as central California (above an elevation of 3500 meters), along the Rocky Mountains to Guatemala (above 4500 meters), and in the Appalachian Mountains to central New England (above 1200 meters).

ENVIRONMENT

Climate

There are many differences between the climates of the High Arctic, Low Arctic, and Alpine tundras, but in general terms they are all characterized by short growing seasons, low temperatures, and strong winds (Figure 7.2).

Arctic.   Arctic tundra, since it occurs at high latitudes, has day lengths that vary greatly throughout the year. North of the Arctic Circle (66° 30′ north latitude) some summer days have 24 hours of sunlight and some winter days have 24 hours of darkness. With very little solar radiation during the winter, temperatures are extremely low. As the days lengthen with the approach of summer, air temperatures do not rise much above the freezing point, since the solar energy for nearly half of the sun season goes to melting snow and thawing the upper levels of the soil. This delay in the onset of warm temperatures is part of a generally late spring that combines with an early winter to give the arctic tundra a short growing season of 60 to 100 days. Even in midsummer, temperatures seldom exceed 15°C, and average temperatures remain relatively cool because of the low angle of the sun. Light intensity is also reduced by the frequently dense cloud cover. However, on clear summer days the Arctic tundra, with the longer day lengths, receives

FIGURE 7.1 Distribution map of the Arctic tundra. The scale of the map is too great to show Alpine tundra.

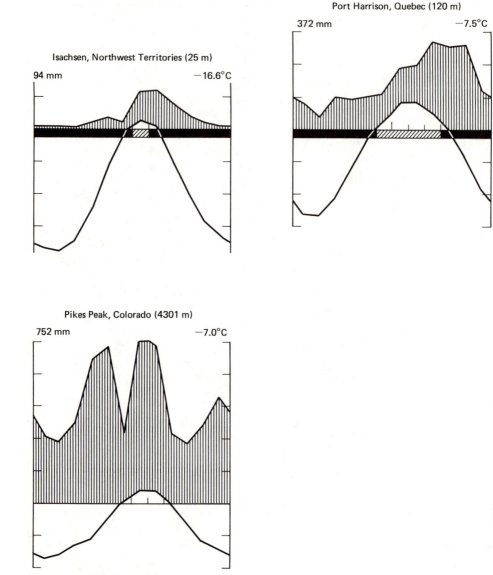

FIGURE 7.2 Climate diagrams for the tundra region: Isachsen has High Arctic tundra vegetation; Port Harrison has Low Arctic tundra vegetation; Pikes Peak has Alpine tundra vegetation. *Sources.* Redrawn from Walter, H. and H. Lieth. 1967. Klimadiagramm-weltatlas. Fischer, Jena, East Germany; Walter, H., E. Harnickell, and D. Mueller-Dombois. 1975. Klimadiagramm-karten. Fischer, Stuttgart, West Germany. 36 p. + maps.

about the same amount of solar radiation per 24-hour period as the Alpine tundra.

The strong winds of the Arctic tundra have the greatest impact during the winter, when gales are frequent and often last for long periods of time. Winds of 65 kilometers per hour over a 24-hour period are not uncommon. These winter winds carry snow and drop it in areas of decreased wind velocity, such as across ridges and in depressions. Even small changes in relief produce drifting snow. The microenvironment of these sites is characterized by a protective layer of snow during the winter, a growing season shortened by the persistence of snow in the spring, and a supply of melt water throughout much of the summer.

One of the lesser known aspects of the Arctic tundra is that it receives relatively little precipitation, especially in the High Arctic where annual totals are commonly smaller than in temperate-zone deserts. However, except for the Polar deserts, most of the Arctic tundra appears to be far wetter than precipitation values would indicate. Reasons for the wet appearance include the spring melt of accumulated snow, cool summer temperatures which slow evaporation, little water runoff because of low topographic relief, and poor soil drainage as a result of permafrost, a permanently frozen layer beneath the soil surface.

Alpine. Except for a short growing season, low average temperatures, and strong winds, the climate of most of the Alpine tundra is different from that of the Arctic. A major reason for this is the smaller variation in day lengths at lower latitudes. Day lengths during the growing season are far shorter in the Alpine tundra than in the Arctic; however, light intensities are much greater in alpine regions because of the higher angle of the sun and the thinner atmosphere. The latter is also responsible for the Alpine tundra's greater exposure to high-energy ultraviolet radiation.

The daily alternation between light and dark in the Alpine tundra causes large fluctuations in temperatures. The intense solar radiation received during the day can raise summer temperatures at the soil surface to over 40°C, but heat loss at night may result in a drop to below the freezing point. Daily temperature variations are greatest during the growing season and increase at lower latitudes. Also, plant temperature variations of 20°C have been observed to occur in a few seconds when clouds pass in front of the sun. This and other abrupt changes in various climatic factors (such as light, wind, and humidity) are characteristic of the Alpine tundra.

Another major difference between the climates of the Alpine and Arctic tundras is the higher precipitation in the alpine zone. Since much of this falls as snow, the start of the growing season is postponed by the time needed to melt it. The length of the growing season is about the same as in the Arctic tundra, but again it varies from site to site, since the high winds cause

FIGURE 7.3 An Alpine tundra and timberline area in mid June in Rocky Mountain National Park, Colorado. The patchy distribution of snow cover results in significant differences between the lengths of growing seasons of individual sites.

snowdrifts which shorten the growing season considerably (Figure 7.3). Despite greater precipitation, the soil during the growing season is generally drier than in the Low Arctic, since soil drainage is better and steeper topographic relief results in higher runoff. Therefore, except for the High Arctic Polar deserts, drought is much more likely in the Alpine than in the Arctic tundra. In fact, late melting snow in the spring combines with drought in the late summer or early fall to produce the Alpine tundra's short growing season.

Soil

Arctic.   The soils of the Arctic tundra may be classified as the Tundra Great Soils Group. Specifically, the Low Arctic has wet peat soils in which gleization forms what may be called a tundra glei soil (page 39), and the High

FIGURE 7.4 Soil profile with permafrost. At this time of year, early June, the surface of the frozen layer at this location is at a depth of 75 cm. *Source.* L. C. Bliss, unpublished.

Arctic has a much drier mineral soil known aś arctic desert soil. A major characteristic of both types is the permafrost layer (Figure 7.4). Depending on the severity of the climate and the thickness of the insulating plant cover, this permanently frozen layer may be up to a few meters thick and in the summer occur as far down as a few decimeters below the surface. The permafrost blocks water drainage and limits root growth but, as its upper surface thaws during the summer, it provides a source of soil moisture—an important factor in dry soils.

A second major characteristic of Arctic tundra soils is that they are unstable. Solifluction—the downslope sliding of water-saturated soil—is very common, especially on top of permafrost. Also, upper soil movement occurs to form polygons, stripes, hummocks, and boils. These patterns develop with the expansion and contraction of water as it freezes and melts. As water freezes and expands against the underlying permafrost, the soil itself is forced upward. The greater the water content of the soil, the greater the expansion. Contraction occurs with extreme winter cold and with the spring thaw. Repeated soil movements eliminate much profile development and bring about an uneven surface. Polygons of up to several meters across form where unequal soil movements sort rock fragments, forcing larger ones to the outside. Some polygons have raised centers (photograph c, page 73). Others have depressed centers that often fill with water in the spring to form numerous small lakes (Figure 7.5). Stone stripes develop similarly but on slopes. Both polygons and stripes result in vegetation patterns coinciding with soil patterns.

Nutrient levels in Arctic tundra soils are generally low. This is the result of several factors, including slow rates of soil formation, decomposition, and plant nutrient uptake (recycling) from the lower soil. Nitrogen is characteristically in low supply, possibly because cool temperatures inhibit bacterial action in the nitrogen cycle. Important inputs of soil nitrogen are supplied by the symbiotic bacteria of mountain avens (*Dryas* spp.) and certain legumes, such as alders (*Alnus* spp.). These plants are capable of increasing soil nitrogen levels by more than 10 times. In most cases, their nodules are just below the soil surface where temperatures are relatively warm. Blue-green algae, some of which may be capable of nitrogen fixation, are often abundant on nitrogen-deficient soil. Animal dung also can be an important source, especially around rocks used as perches.

Alpine.   There are a wide variety of soils in the Alpine tundra, but all are young with a high-rock content. Fell-fields are widespread. Permafrost

FIGURE 7.5 Aerial view of large depressed center polygons on Melville Island, Northwest Territories. *Source.* L. C. Bliss, unpublished.

occurs infrequently. Surface patterns such as polygons form in some areas but are not as common as in the Arctic tundra. Solifluction, although accelerated by steeper relief, is less widespread because of the dry, shallow, rocky character of the soil and the lack of permafrost. Nutrients are usually at low levels, and, as in arctic soils, the breakdown of parent material is important in their availability.

VEGETATION

Floras

Tundra vegetation evolved not as a direct product of any of the three Tertiary geofloras, but instead as a response to major changes in relief and climate during the Tertiary Period (Chapter 6). Tundra vegetation first appeared in the late Tertiary; the flora may be much older. The flora was produced by numerous species invasions, rather than by a few invasions followed by large diversification. Evidence of this is that there are few genera restricted to the tundra, but a large number of species and an abundance of subspecies. The species that invaded the tundra environments are thought to have been preadapted, in the sense that some of their morphological and physiological characteristics were already suited for growth and reproduction there. Coastal, bog, and marsh habitats of cool-temperate regions were likely sources of these species. Phenotypic plasticity—the ability of individuals to adjust to new or changing environments—was and remains of great importance to tundra plants.

The floras of the North American Arctic and Alpine tundras are not as diverse as the floras of temperate and tropical regions. In fact, the vascular flora of the Arctic Slope of North America consists of less than 500 species, subspecies, and varieties. One striking aspect of the tundra floras is that many species have a very wide distribution. Numerous Arctic tundra species are circumboreal, occurring in North America and Eurasia. Many are also found in the Alpine tundra, implying a common origin of the two floras. Approximately 45 percent of the species of a Rocky Mountain tundra in Montana also are found in the arctic. This figure is 75 percent for an Appalachian Mountain tundra in New Hampshire, and 15 to 20 percent for the Sierra Nevada Alpine tundra flora in general. However, there are significant differences between the floras of the three North American alpine regions; that of the Sierra Nevada is especially distinct.

Since individual tundra species frequently have a wide distribution over a range of somewhat different environments, local populations may differ from each other morphologically, physiologically, and genetically. This is the result of adaptation of each population to its local environment. Such populations are considered **ecotypes**—ecological races of the same species. Moun-

FIGURE 7.6   Mountain sorrel in Sequoia National Park, California.

tain sorrel *(Oxyria digyna)* is a well-studied example of a tundra species made up of ecotypes (Figure 7.6; see also H. A. Mooney and W. D. Billings, Suggested Readings, this chapter).

Communities

Arctic.   The Low Arctic is floristically and vegetationally more diverse than the High Arctic, probably because of greater precipitation and a wider range of soil-moisture conditions. The region of the Low Arctic just north of the Boreal coniferous forest is characterized by Low Shrub tundra dominated by willows and birches *(Betula* spp.) of up to a meter in height (Figure 7.7). The understory consists of dwarf heath shrubs, cottongrass *(Eriophorum* spp.), sedges, mosses, lichens (especially reindeer lichens, *Cladonia* spp.), and a variety of forbs. Many of the understory species extend southward into the transitional region between the tundra and the forest. The Low Arctic also has Tall Shrub communities of one to three meters height. These are common along rivers and streams and on well-drained slopes. Dominants include willows, birches, and alders. The hilly uplands of the Low Arctic have a variety of Herb and Dwarf Shrub communities, with cottongrass, sedges, mosses, lichens, and dwarf heaths as the most common plants (Figure 7.8). Lastly, the lowlands and coastal plain regions are marshy areas of sedges, cottongrasses, and true grasses (Figure 7.9).

In the High Arctic, areas of greatest soil moisture, such as some coastal

FIGURE 7.7 Low Shrub tundra vegetation with dwarf birch (*Betula nana* ssp. *exilis*), willows, heaths, lichens, and mosses at Eskimo Lakes, Northwest Territories. *Source.* L. C. Bliss, unpublished.

FIGURE 7.8 Dwarf Shrub tundra vegetation with heaths and lichens on raised polygons in the Mackenzie Delta, Northwest Territories. A sedge (*Carex aquatilis*) appears in bands along areas of water where ice wedges form in winter. *Source.* L. C. Bliss, unpublished.

FIGURE 7.9 Wet sedge tundra vegetation dominated by cottongrasses, a sedge *(Carex aquatilis)*, and a grass *(Dupontia fisheri)* in late June on the coastal plain near Prudhoe Bay, Alaska. There is standing water in the foreground. *Source.* L. C. Bliss, unpublished.

lowlands and valley bottoms, have communities that resemble those of the Low Arctic, including Herb and Dwarf Shrub communities. These areas have a vegetation cover of 80 to 100 percent, but such habitats cover little total area. Polar semideserts are much more widespread; they have a vegetation cover of only 5 to 20 percent, except where extensive moss and lichen growth increases it to 50 percent (Figure 7.10). Semideserts are found from sea level to 100 to 200 meters elevation. Polar deserts occur at higher elevations, as well as down to sea level in those regions where the polar ice pack lasts throughout the year. Their cover of vascular plants is frequently 0 and is otherwise less than 2 percent, though mosses and lichens can raise total cover to 20 percent. Greatest cover occurs around snowdrifts where melting snow provides soil moisture and adjacent to rocks where animals perch and enrich the soil.

Alpine. The plant communities of the Alpine tundra of North America are highly varied, reflecting differences between regional environments of the major mountain chains and diversity in local habitats. One plant community found in most alpine regions is the Wet Meadow (Figure 7.11). It is dominated by a dense mat of sedges, with some grasses and scattered forbs. Wet Meadow vegetation occurs on sites that remain moist throughout the growing season, such as stream banks and flats that are downslope from melting snowbanks. A second community type is the Dry Meadow. It also is charac-

FIGURE 7.10 Polar semidesert vegetation dominated by various forbs, mosses, and lichens in Ellef Ringes Land, Northwest Territories. *Source.* L. C. Bliss, unpublished.

FIGURE 7.11 Wet Meadow Alpine tundra vegetation dominated by various sedges in Sequoia National Park, California.

FIGURE 7.12   Dry Meadow Alpine tundra vegetation dominated by various sedges, mosses, and lichens in Rocky Mountain National Park, Colorado.

terized by sedges with some grasses and forbs, but it is present on sites that become dry in the latter half of the summer (Figure 7.12). Alpine fell-field vegetation is composed of forbs, sedges, and grasses that are between rocks (Figures 7.6 and 7.14). Lichens and mosses may also be on the rocks. Sometimes dwarf heath and willow shrubs grow in fell-fields, appearing as a low matlike growth-form in windswept areas. Tall Shrub communities dominated by willows may occur along streams.

Production

Net primary productivity in tundra vegetation tends to be very low, because it is influenced by factors such as cool temperatures, low soil-nutrient levels, and short growing seasons. Net productivity usually ranges from 100 to 400 grams of dry weight per square meter per year, but may be far less under extreme conditions. For example, values of 9 grams per square meter per year have been recorded for the High Arctic. Only vegetation in the most severe temperate deserts may have similarly low values. However, one problem in comparing the annual productivity of the tundra with that of other formations is differences in the length of growing seasons. Calculations of net primary productivity values on a per day basis during the growing season result in smaller differences between the tundra and other types of vegetation. Two additional problems with net production comparisons are (1) a great proportion of tundra plant biomass is below ground, and this cannot be excluded from calculations, as is often done; and (2) measurement of net production by biomass alone may not be accurate, since at least in the case of alpine species their tissues have higher caloric (energy) values than those of many temperate and tropical species, indicating a higher amount of production per unit weight of dry matter.

Adaptations

Wind.   Winter winds are one of the more important environmental factors in the tundra. They have two major, damaging effects on the exposed parts of plants: one is abrasion by wind-borne snow and ice crystals; the other is desiccation by increased transpiration at a time when water cannot be absorbed from the frozen soil.

Many tundra species are restricted to areas where drifting snow quickly covers and protects them in the early winter. Some snowbank species may absorb melt water through their leaves, replacing water which was lost before snow covered the plant.

Species not restricted to snowbank sites have other adaptations to wind. The short or dwarf stature of most tundra plants increases the likelihood that they will be at least partially covered with snow (Figure 7.13). Also, wind

FIGURE 7.13   Extreme dwarfness in tundra plants. A mature, flowering shrub of bearberry willow *(Salix uva-ursi)* is on the right side of the photograph. Several lichens and mosses cover the rock to the left side of the shoe. Location: Mt. Washington, New Hampshire.

FIGURE 7.14   The pillowlike cushion growth-form of diapensia *(Diapensia lappon-ica)* appears in the center of the photograph. This fell-field area also has dwarf shrubs, a dwarf tree, and sedges. Location: Mt. Washington, New Hampshire.

speed is slower close to the ground. Dwarfness in shrubs is often genetic, but it also may be caused by the death of exposed upper buds. Persistent dead leaves and stems are adaptive, since they take the brunt of wind abrasion. They also slow wind flow and create eddies that cause blowing snow and ice particles to drop and build up around the plant. Some species have such dense clusters of stems that they are pillowlike in appearance (Figure 7.14). Inside of these "cushion" growth-forms, wind velocities are reduced by up to 99 percent.

Other adaptations include the ability of some alpine species to survive winter drought. Alpine azalea *(Loiseleuria procumbens)*, for example, produces adventitious roots late in the winter and these absorb melt water on the surface of the still frozen soil. Other species may develop winter drought resistance, perhaps by the conversion of stored starch to soluble sugars. Some winter drought adaptations are also important for survival of summer drought.

Winter winds, while largely a problem for tundra plants, also aid in their dispersal—a factor of importance since tundra species generally lack the varied dispersal mechanisms of temperate plants. During the winter, fruits and seeds are blown across the hard-packed snow for great distances. A few tundra plants have the means to release their seeds gradually and only during strong winter winds.

Growing Season.   Another major problem for the growth of plants in the tundra is the short growing season of 60 to 100 days. This environmental factor explains why 98 to 99 percent of all tundra plants are perennials. A few annuals occur in the most favorable sites, but they remain very small, conserving their yield from photosynthesis and using it for rapid reproduction (as do desert ephemerals, page 200). Perennials, however, can survive without reproducing each year; they only require a net production that is sufficient for initial growth the following spring. Another advantage of perennials is that they do not lose valuable days out of the short growing season for the growth of entirely new vegetative bodies, as do annuals. The evergreen habit would seem to be advantageous in this same sense, if it were not for problems with winter winds. However, many plants are semievergreen, in that they have a few leaves or parts of leaves that remain green and are capable of photosynthesis when first exposed to light in the spring. Examples of semievergreen plants are the very widespread sedges, grasses, and other monocots, whose basal parts of the inner leaves remain green throughout the winter and are pushed above the protecting old dead leaves with the first growth in the spring.

In addition to growth-forms, tundra species also possess adaptive growth patterns. For example, some tundra plants opportunistically use as much of the growing season as possible, not ending their growth at any particular stage of development; they continue growth until the weather changes in the fall and then resume the following spring. These aperiodic species are thus able to take advantage of occasional longer growing seasons, but in the fall they are susceptible to injury from sudden changes in the weather. Most aperiodic species are found in snowbank sites, where growing seasons may be irregular in length, and where early protecting snow cover is likely. Except for snowbank species, the majority of tundra plants exhibit a more conservative growth pattern—periodic growth—in which plant development is stopped at a particular stage in late summer, regardless of weather conditions. This eliminates the possibility of additional growth in mild years but helps ensure against injury in the fall.

Both aperiodic and periodic tundra perennials typically have a spurt of growth immediately after the snow melts. Their respiration greatly exceeds photosynthesis until 75 to 90 percent of shoot growth is completed, usually within two weeks. This rapid growth is especially beneficial for deciduous plants, since it allows them to quickly form new sets of leaves.

In most perennials, the energy for spring growth comes from carbohydrates stored in below-ground organs, such as roots and bulbs. Tundra perennials typically have over 50 percent of their biomass below ground and in some cases up to 95 percent. The carbohydrates lost from storage are replaced over the rest of the summer when photosynthesis exceeds respiration. Usually, one month of good weather is enough for accumulation of storage materials sufficient for growth the next spring. Again, the short stature of tundra plants may be considered an adaptation, since carbohydrates are not used up in nonessential growth but are conserved. In addition, some herbaceous tundra perennials transfer organic compounds from their aerial portions to their storage organs before winter; therefore, little more than cellulose cell walls are lost when the aerial parts die.

The short growing season also affects various aspects of reproduction. Again, the conservation of carbohydrates is extremely important. Of course, aperiodic species may stop reproductive development at any stage and resume the following spring, but even periodic species may spread reproduction over two years and thus not expend large amounts of carbohydrates in any single growing season. Also, some species have a type of feedback control, in that low carbohydrate supplies may inhibit flowering. When seeds are produced, they are usually very small and lack accessories, such as fleshy or burred fruits.

Many tundra plants, especially arctic species, are capable of self-pollination, unlike most temperate plants. Apparently, arctic species cannot afford any delay in reproduction that obligate (mandatory) out-crossing might produce. Cross-pollination is more common in the Low Arctic and in regions where large insect populations help ensure against any long delay in pollination.

Vegetative reproduction is very common in the tundra; it is less complex than sexual reproduction and, hence, is adaptive in areas with severe climate. Many tundra species, especially sedges and grasses, reproduce primarily by rhizomes, stems which spread underground and intermittently sprout aerial shoots. Some arctic-alpine species, such as mountain sorrel (Figure 7.6), form rhizomes in the arctic but not in the alpine, indicating ecotypic variation within the species. Layering—the rooting of plants where branches touch the ground—is common in cushion plants and prostrate shrubs.

In most situations, asexual reproduction is thought to be evolutionarily disadvantageous; no genetic recombination takes place, and with little genetic variation, environmental change can result in the local extinction of a species. However, in extreme environments such as the tundra, asexual reproduction has the advantage that individual gene combinations, which are adaptive to the environment, are not lost through out-crossing, but are maintained, since offspring have the same sets of genes as the parents. The evolutionary disadvantage of little genetic variation is not as important in the tundra as it is elsewhere. The tundra environment is dominated by physical factors, such as the climate. Environments such as this are relatively consistent and do not require continual readaptation as is necessitated by competition in environments where biological interactions are more important. Also, the high degree of phenotypic plasticity in tundra plants aids their survival of variations in climate. In the case of many tundra species, asexual reproduction is common in extreme environments, and sexual reproduction may be typical for the same species under more mild conditions.

Temperature. Relatively low tempeatures during the growing season usually reduce rates of metabolic activity, but many tundra plants have compensating adaptations. Any plant characteristic that results in reduced wind flow—such as short height, the cushion growth-form, and persistent dead leaves—is adaptive, since the rate of heat loss is decreased and higher metabolism ensues. Short height also enables plants to metabolize in a microenvironment that is warmed by the radiation of energy absorbed by the soil surface. An excellent example of this is the crustose lichen, which grows in more extreme climates than the seed plant. Other examples are rosette plants that flower before elongating and shrubs such as the alder that leaf out at their bases several days earlier than at their tops.

Plant temperatures may be increased by pigments—the darker the color, the more sunlight is absorbed. Some tundra plants metabolize large amounts of anthocyanin, a red pigment which in combination with chlorophyll may give a nearly black color. In one arctic species it has been shown that this pigmentation combined with dense growth resulted in a temperature of 15.5°C over the outside air temperature. One of the most interesting examples of the advantages of dark pigmentation is demonstrated by certain arctic snowbank species, which absorb light that passes through the snow and give off heat that is trapped and retained around the plants. These plants actually photosynthesize beneath the snow, and by this mechanism lengthen the short growing season of the snowbank site by two weeks.

Plants with a dense covering of clear epidermal hairs may also make use of a "greenhouse effect." In addition, such hairs could be important in reducing heat loss at night in alpine species.

Regardless of these mechanisms, the temperatures of most tundra plants are still below those that would be considered normal in temperate climates. Consequently, tundra plants are physiologically adapted to carrying on their metabolic activities at low temperatures. For example, most tundra plants have a lower optimum temperature for photosynthesis than temperate plants. In fact, at least in mountain sorrel and arctic meadow-rue (*Thalictrum alpinum*), the optimum temperature is usually lower for arctic populations than for alpine populations. Other studies have shown that some alpine individuals have a lower absolute-minimum temperature for photosynthesis than lowland plants of the same species. It has been reported that photosynthesis can be carried on by some tundra species, including the mountain sorrel and a species of buttercup (*Ranunculus glacialis*), at below-freezing temperatures.

Dark (night measured) respiration rates are higher for arctic individuals than for alpine individuals of the same species, and rates for tundra plants are higher than for others kept at the same temperature. Rates of both respiration and photosynthesis are thought to be partly environmentally conditioned; thus, metabolic capacities may vary throughout the year. Such phenotypic adjustment is believed to maximize efficiency.

At the same time that they carry on normal metabolism, tundra plants have to be continually frost-hardy and be able to recover very quickly after a frost. Mosses and lichens are especially capable in this regard. Winter frost resistance is obviously a necessity, but the level of resistance may vary over the winter. For some alpine plants, resistance is at a maximum in early winter, when the chances of being covered by snow are small.

Light.   There are large differences in light intensity between the arctic and alpine environments, and this accounts for certain differences in tundra plants. Arctic plants have leaf areas that are large for the overall size of the plants, an adaptation to weak light intensities. Alpine plants such as mountain sorrel and downy oat-grass *(Trisetum spicatum)* require greater light intensities to reach maximum photosynthesis than do arctic plants, even of the same species, and temperate lowland plants. Also, alpine ecotypes, at least in the case of mountain sorrel and arctic meadow-rue, have a lower chlorophyll content than arctic ecotypes. This results in greater light reflectance by alpine plants—a factor that helps maintain proper leaf-heat balance. Protection from the high light intensities also may be provided by dense, radiation-absorbing epidermal hairs, anthocyanin pigments, and several layers of leaf photosynthetic tissue.

Tundra plants also show adaptations to the differences in day lengths between the arctic and the alpine. The initiation and breaking of dormancy, summer frost hardiness levels, and flowering are thought to be at least partially determined in many tundra plants by the length of the photoperiod. For example, it has been shown that alpine ecotypes of mountain sorrel will flower under photoperiods of 15 hours, but arctic ecotypes of the same species require over 20 hours. The exact photoperiod requirement may even differ somewhat between individuals of a single population. This is thought to be an adaptation to varying snow persistence, since no matter when the snow melts, some individuals will still flower.

Drought.   Adaptations to dry conditions are important in many areas of the tundra, but especially in the alpine, where summer drought is very likely. Epidermal hairs, which are more common on alpine plants than arctic plants, may reduce transpiration by absorbing and reflecting light and by decreasing wind flow near stomates. Many plants of open alpine areas can reduce transpiration by closing their stomates during midday; yet, snowbank species have no such mechanism. Also, plants growing in dry alpine sites are known to be able to photosynthesize at lower water-content levels than plants growing in wet sites. Both mosses and lichens are well suited for periodic drought, since their growth is opportunistic. In the case of lichens, they maintain a very low rate of metabolism when desiccated, and some reportedly can recover to full activity within one minute after becoming moist.

Dormancy. Adaptations related to winter dormancy are a necessity in the tundra. Differences between aperiodic and periodic plants have already been discussed (page 89). In both types of species, the onset of dormancy may be cued by several factors, including shortening day lengths, decreasing soil and air temperatures, increasing drought, and accumulating carbohydrate reserves. Only the snowbank species do not respond and develop high winter hardiness levels. Yet, their dormancy (and that of other species) can be so complete that they can survive at least several years if covered with snow. Lichens are especially capable of this, as are the seeds of some species.

The breaking of dormancy is critical in the tundra, since any delay would further decrease the already short growth period. Some of the factors that may trigger the end of dormancy are melt water, increasing day lengths, rising soil and air temperatures, and completion of a lengthy cold period. The requirements for seed germination are especially exacting. For example, some seeds require relatively high spring temperatures. This ensures that germination occurs only in mild years, when it is more likely that the plant will reach maturity before fall. The development of many seedlings in their first season is characterized by rapid root growth and by correspondingly slow shoot growth—an advantage in locations with a late summer drought. Nevertheless, successful seedling establishment is rare.

Human Impact

Because of its climate and low net production, the tundra was not utilized to a significant extent in presettlement times. In fact, the tundra can be said to have evolved without human influence. In the arctic, the North American Inuit (Eskimo) has always had low population densities and has made more use of the seas than the tundra. Alpine regions were utilized even less. Small Indian campsites have been found but, since these are scattered along old trade routes, they were probably used infrequently for brief stays.

Modern times have brought a greater human impact on the tundra; yet, it has been relatively small in comparison to that on other vegetation formations. Perhaps the greatest change in the Alpine tundra has been the destruction of vegetation cover and subsequent erosion brought about by the grazing of domestic livestock, particularly sheep. Such grazing was a greater problem in the past, before the establishment of federal parks and forests eliminated or restricted it. More recently, there has been an increase in damage from the recreational use of the alpine zone. For example, hiking trails concentrate water runoff and thereby create erosion problems.

The Arctic tundra, with its far greater land area, has been considered as a site for agriculture. Cool temperatures and short growing seasons create obvious difficulties for raising temperate crops, but attempts to raise grazing animals also encounter problems. Even if the production of native plants could be increased by fertilization or other means, much of the growth would not be available for grazers, since it would be below ground. In any case, increased grazing would greatly alter the arctic ecosystem, since the storage of carbohydrates is essential for the continuing survival of the perennial plants.

Future human impact on the Arctic tundra will come from activities associated with the removal of mineral resources. A widely publicized example is the transport of oil from northern Alaska through a pipeline to an ice-free port on the state's southern coastline. Precautions were taken in the pipeline design to reduce interference with migratory wildlife populations and to avoid melting the permafrost (so that the pipeline would not sink and rupture). It will be many years before the full impact of this project is known but, without doubt, it represents only the first of what will be repeated attempts to remove mineral reserves from the Arctic tundra.

Except for problems unique to individual resources, the nature and extent of the damage done to the tundra greatly depends on the vegetation and the soil. Although some tundra vegetation recovers following disturbance, in most cases secondary successions proceed slowly, especially if soil erosion has occurred. A complicating factor is that, when the vegetation is disturbed, part of the insulation for the permafrost is lost, and it can begin to melt, forming a depression which becomes a site for water runoff and erosion. This exposes additional permafrost, compounding the problem. In many cases, merely the weight of a vehicle moving across tundra during the summer is enough to start this process. Therefore, unless specially designed equipment is used, construction activity must take place during the winter when the ground is frozen; however, this is when the weather is most severe.

SUGGESTED READINGS FOR FURTHER STUDY

Billings, W. D. 1973. Arctic and alpine vegetations: Similarities, differences, and susceptibility to disturbance. BioScience 23:697–704.

Billings, W. D. and H. A. Mooney. 1968. The ecology of arctic and alpine plants. Biological Reviews 43:481–529.

Bliss, L. C. 1956. A comparison of plant development in microenvironments of arctic and alpine tundras. Ecological Monographs 26:303–337.

————. 1962. Adaptations of arctic and alpine plants to environmental conditions. Arctic 15: 117–144.

————. 1963. Alpine plant communities of the Presidential Range, New Hampshire. Ecology 44:678–697.

————. 1971. Arctic and alpine plant life cycles. Annual Review of Ecology and Systematics 2:405–438.

———— (ed.). 1978. Truelove Lowland, Devon Island, Canada: A high arctic ecosystem. University of Alberta Press, Edmonton. 735 p.

Bliss, L. C., G. M. Courtin, D. L. Pattie, R. R. Riewe, D. W. A. Whitfield, and P. Widder. 1973. Arctic tundra ecosystems. Annual Review of Ecology and Systematics 4:359–399.

Britton, M. E. 1957. Vegetation of the arctic tundra, p. 22–61. *In* H. P. Hansen (ed.). Arctic biology. Proceedings, 18th Annual Biology Colloquium. Oregon State University Press, Corvallis.

Hansen, H. P. (ed.). 1957. Arctic biology. Proceedings, 18th Annual Biology Colloquium. Oregon State University Press, Corvallis. 318 p.

Hanson, H. C. 1953. Vegetation types in northwestern Alaska and comparison with communities in other arctic regions. Ecology 34:111–140.

Ives, J. D. and R. G. Barry (eds.). 1974. Arctic and alpine environments. Methuen, London, 999 p.

Johnson, P. L. and W. D. Billings. 1962. The alpine vegetation of the Beartooth Plateau in relation to cryopedogenic processes and patterns. Ecological Monographs 32:105–135.

Major, J. and D. W. Taylor. 1977. Alpine, p. 601–675. *In* M. G. Barbour and J. Major (eds.). Terrestrial vegetation of California. Wiley, New York.

Mooney, H. A. and W. D. Billings. 1961. Comparative physiological ecology of arctic and alpine populations of *Oxyria digyna*. Ecological Monographs 31:1–29.

Mooney, H. A. and A. W. Johnson. 1965. Comparative physiological ecology of an arctic and an alpine population of *Thalictrum alpinum* L. Ecology 46:721–727.

Savile, D. B. O. 1972. Arctic adaptations in plants. Monograph No. 6. Research Branch, Canada Department of Agriculture. 81 p.

Tranquillini, W. 1964. The physiology of plants at high altitudes. Annual review of Plant Physiology 15:345–362.

8

CONIFEROUS
FOREST VEGETATION

Coniferous forests are common in the northern hemisphere. In North America this formation occurs across Canada and Alaska, through the major mountain chains, and along the Pacific Coast as far south as central California (Figure 8.1). These forests are dominated by species of the conifer section of the gymnosperms, especially species of pines (*Pinus* spp.), spruces (*Picea* spp.), and firs (*Abies* spp.). Typically, the dominants have a cone-shaped growth-form. The vast majority of conifers are evergreen, but there are significant exceptions. The leaves are generally needlelike in appearance and quite strong; botanically, they are narrow-sclerophylls. The height of the dominants is highly variable; conifers in central Canada are a few meters tall, and some in northwest California exceed 100 meters. Most coniferous forests have a vertical stratification of four layers: surface, herb, shrub, and canopy. A few also have a subcanopy layer. Regardless of this variation, the different coniferous forests have a common origin. All were derived from the Arcto-Tertiary Geoflora as it was modified by continental cooling and drying trends, mountain uplifting, and glaciation (Chapter 6).

BOREAL CONIFEROUS FOREST

Distribution

The Boreal coniferous forest forms a wide band south of and parallel to the Arctic tundra in North America and Eurasia. Treeline—its boundary with the Arctic tundra—is quite irregular; the forest may extend into the tundra

(a) Boreal coniferous forest stand in southcentral Ontario.
(b) A winter scene in a young Rocky Mountain coniferous forest in northcentral Colorado.

(c) A montane meadow and forest in Sequoia National Park, California.
(d) Early morning in a mature stand of the Northwest Coastal coniferous forest in northern California.

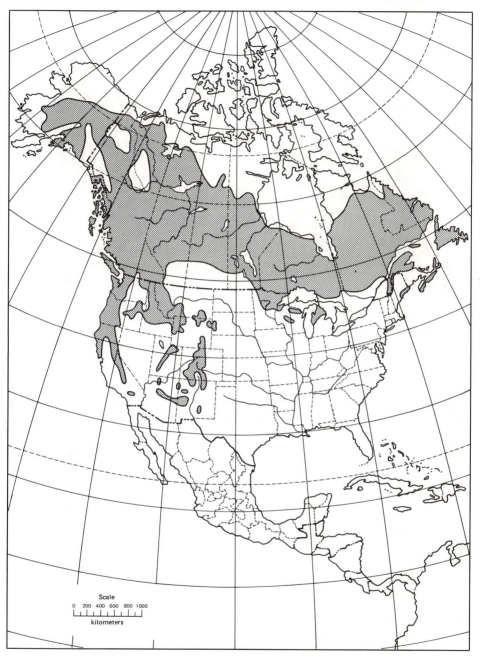

FIGURE 8.1 Distribution map of the coniferous forest. The scale of the map is too great to show the full range of the Appalachian coniferous forest.

FIGURE 8.2 Treeline, the ecotone between the Arctic tundra and the Boreal coniferous forest, east of Inuvik, Northwest Territories. Despite the small size of the black
spruce, these trees may be quite old. Shrub and herb species include dwarf birch
(*Betula nana* ssp. *exilis*), a willow (*Salix pulchra*), a cottongrass (*Eriophorum vagina-
tum*), true grasses, and lichens. *Source.* L. C. Bliss, unpublished.

for distances of up to 350 kilometers in sheltered valleys, and tundra vegetation occurs well into the Boreal forest region along exposed ridges. Treeline,
where trees are short and may be knarled and twisted, is dynamic, shifting
continually and slowly in response to long-term environmental changes
(Figure 8.2). The southern boundary of the North American Boreal forest is
obscure in most areas. In the east it blends with the Hemlock-White Pine-
Northern Hardwoods forest, a transitional association of the deciduous
forest (page 142). In southcentral Canada the Boreal forest zone is dominated
by deciduous trees that intergrade into the grassland formation. In the west
there is a gradual transition from the Boreal forest into the Rocky Mountain
and the Northwest Coastal coniferous forests.

Environment

The environment of the Boreal forest region resembles that of the Arctic
tundra. The climate is characterized by low temperatures, low precipitation,
and short growing seasons that range from 90 to 120 days (Figure 8.3).
Although day lengths are long during the growing season, the low angle of
incidence of solar radiation keeps temperatures cool. Peak temperatures

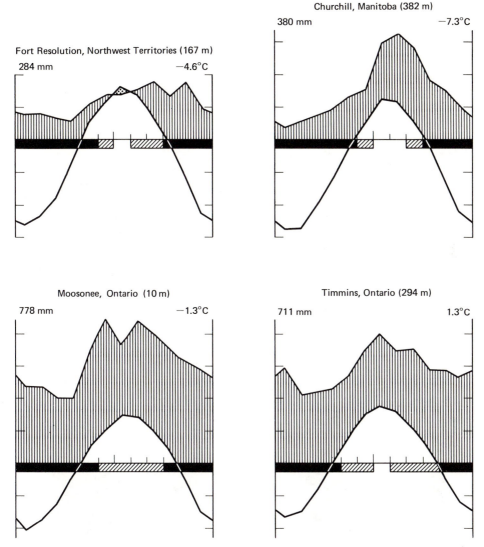

FIGURE 8.3 Climate diagrams for the Boreal coniferous forest region: Fort Resolution is the furthest north and west of these locations; Churchill is close to treeline near Hudson Bay; Moosonee is furthest east; Timmins is furthest south. *Source.* Redrawn from Walter, H. and H. Lieth. 1967. Klimadiagramm-weltatlas. Fischer, Jena, East Germany.

during the summer are usually less than 30°C, and in some areas frosts are always a possibility. Short day lengths during the winter months result in very low temperatures; during the coldest month mean daily minima are commonly near −15°C. Annual precipitation in the central portion of the Boreal forest averages only around 40 to 50 centimeters. Nevertheless, the region appears quite wet; much of the landscape is dotted with lakes and bogs, and soil moisture is frequently high. The wet appearance is partly explained by the cool summer temperatures that result in low evaporation rates.

The Great Soil type of the Boreal forest is the Podzol. It is often poorly drained, but still well leached and consequently nutrient poor and quite acidic. The soil profile consists of a dark organic layer above a greyish-white sandy band. The organic layer persists, because decomposition is slowed by cool temperatures and short growing seasons. The decaying organic matter is highly acidic; this increases both the leaching and the acidity of the other horizons. Permafrost is sometimes present, especially in bog areas. In much of the Boreal forest region the soils are thin, having developed on bedrock exposed by glaciation, but they are thicker in areas of glacial deposits, especially to the south.

The topographic relief of the Boreal forest region is largely a product of extensive past glaciation. In general, the relief is low with numerous depressions that are filled with lakes and bogs (Figure 8.4). The flat topography limits water runoff and thus adds to the wet aspect of much of the region.

FIGURE 8.4 A view of a typical Boreal coniferous forest landscape east of Lake Superior in Ontario. Dominant tree species include white spruce, balsam fir, and birches.

Fires are another important part of the Boreal forest environment. Periodically large areas are burned, especially in drought years when soil and bog surfaces become dry, making the organic matter highly combustible.

Vegetation

Structure and Composition. The trees of the Boreal forest are small in comparison to those of other coniferous forests. They are usually under 15 meters in height and less than 0.5 meters in diameter. In the Taiga—the northern third of the forest—the trees range from 4 meters in height to a short, shrubby form at treeline (Figure 8.2). Tree density is high in wet areas (photograph *a*, page 97) but, toward the north and in dry sites, the forest tends to open up. The shrub layer is also denser in wet areas, especially along rivers. The herb layer is generally not as well developed as it is in the deciduous forest to the southeast. Several reasons for this have been suggested; undoubtedly, seedling establishment is difficult with continual shade and a thick organic layer. Lastly, the surface layer is usually very well developed; it covers tree trunks, limbs, and, in many areas, much of the forest floor. Mosses tend to predominate in wet areas, and lichens are more common in drier habitats.

The Boreal forest has the lowest species diversity of any major North American forest type. In fact, of all the formations, only the tundra is poorer in the number of species present. Greatest species diversity is found in the southeast part of the forest. Floristically, the North American Boreal forest is closely related to the still less diverse Eurasian Boreal forest; although they have no dominant tree species in common, the same genera occur in both. The floras of the North American Boreal and high-elevation Montane coniferous forests are even more closely related.

The Boreal forest is generally a spruce-fir forest. The most widespread North American dominants are white spruce *(Picea glauca)*, black spruce *(Picea mariana)*, and balsam fir *(Abies balsamea)*. Black spruce and tamarack or larch *(Larix laricina)*, a deciduous conifer, are especially common in bogs to the south (Figure 8.5), but in the north may be found on drier sites near or at treeline. Jack pine *(Pinus banksiana)* and two deciduous angiosperms, paper birch *(Betula papyrifera)* and quaking aspen *(Populus tremuloides)*, may dominate stands following a disturbance, such as fire; however, some of their stands are considered to be mature or climax in certain habitats. Quaking aspen is also the dominant in the "aspen parklands," the transitional region between the Boreal forest and the northern grasslands.

Most of the shrubs are willows *(Salix* spp.) or heaths (members of the Ericaceae). Many of the herbs are also heaths. The dominant lichen genus is the same as that of the tundra, the reindeer lichen *(Cladonia* spp.). A common moss is peat moss *(Sphagnum* spp.); it is especially important in the forma-

FIGURE 8.5 The edge of a boggy area in northeast Minnesota. The major tree species are tamarack and black spruce, with white spruce to the left.

tion of bogs. At times peat moss produces a thick mat that floats on the surface of lakes. These mats can support the weight of humans, whose walking causes the surface to shake. Over a period of many years, the sphagnum can completely cover and fill a lake, producing a bog. Bogs are typically poor in nitrogen and consequently have an interesting flora that includes insectivorous plants (plants which obtain nitrogen from the digestion of insects). Further succession can lead to the establishment of a forest community where originally there had been an open lake (Figure 8.6).

Production.   Net primary productivity values for the Boreal forest tend to be quite low, lower in fact than for any other North American forest. An average would be around 500 grams dry weight per square meter per year; however, as with the tundra, daily net production figures during the growing season are comparable to some types of temperate vegetation.

Adaptations.   Of course, species are adapted to the environment as a whole, but a key factor in explaining the adaptations of Boreal forest species is the short growing season. One adaptation exhibited by the vast majority of the dominant tree species and heath shrubs is evergreenness—the retention of most leaves throughout the year. Evergreen species are able to photosynthesize as soon as environmental conditions permit it in the spring; this enables them to make use of the entire growing season, rather than delay maximum photosynthesis until a new set of leaves has been formed. Also, evergreens

FIGURE 8.6 A boggy area in northeast Minnesota where succession is leading to the development of a typical forest stand of the same species shown in Figure 8.5.

can afford to postpone growth until the fluctuating spring weather has stabilized, but deciduous species cannot and consequently could suffer frost damage. The evergreen state is further advantageous in that energy reserves are not reduced each spring for the production of complete, new sets of leaves.

One of the problems for evergreen species is leaf survival during the winter. The leaves have several adaptations to xeric conditions, including a thick cuticle, a strong epidermis, and sunken stomates. Although these morphological characters are of little value during the growing season, they, along with increased sugar content of the sap, are important for surviving winter, when soil moisture is frozen and desiccation is possible. Deciduous trees are obviously adapted to winter drought, since only bark-covered branches and trunks are exposed. In spite of their disadvantage of having to produce entire sets of leaves each spring, the wide distribution of the paper birch, quaking aspen, and tamarack make it clear that the deciduous growth-form is an alternative strategy of adaptation to the Boreal forest environment.

Survival during the winter months is aided by dormancy, when respiration may approach zero. The development of dormancy in Boreal and other species exposed to cold involves hardening—a process by which plants become greatly resistant to frost. Before hardening occurs, the foliage of many evergreen species is susceptible to temperatures of a few degrees below freezing, but after hardening it may survive the low temperature extremes typical of the region.

The soil is another major environmental factor to consider when discussing the adaptations of boreal species. The shallow root system of most conifers is adaptive to habitats where the soil is thin because of bedrock or permafrost. A shallow root system also allows the absorption of water early in the spring, as soon as the topsoil thaws. Many, if not most, conifers and some heaths are mycorrhizal (page 44). The fungi, in effect, enlarge the root systems of the plants and also enzymatically break down organic matter, making nutrients more available to the roots. This is especially important in the Boreal forest, because cool temperatures retard decomposition and the soil is poor in nutrients. Although conifers and heaths have a wide distribution in North America, their dominance in the Boreal forest is striking, especially when one considers that many plant families common elsewhere are not well represented here. The presence of several species of saprophytes—nonphotosynthetic plants which derive their nutrition from the accelerated decay of dead organic matter—also may be indicative of soil conditions.

The seral species of the Boreal forest are well adapted to their role in succession. For example, the small, winged fruits of paper birch and jack pine and the hairy seeds of quaking aspen are adaptive in that they are effectively dispersed by wind into recently disturbed sites. In addition, quaking aspen can quickly regenerate following fire or severe wind storms by sprouting new aerial stems from old root systems. Indeed, sprouting is the most common means of reproduction within this species. Jack pine also can regenerate relatively quickly in an area following fire; many of its seed cones are serotinous; that is, they remain closed and retain seeds until exposed to the high temperatures of a fire. At that time, the sap that seals the cones melts, and the cone scales slowly spread apart. The seeds land on an ash-covered mineral soil, an ideal seedbed for most coniferous species.

Treeline.   The above discussion of the Boreal forest adaptations could be expanded by repeating many of the adaptive characters of tundra herbs described in Chapter 7. However, it is the tree growth-form that is the most striking difference between the Boreal forest and the Arctic tundra; the northern boundary of the Boreal forest is treeline, the limit of tree growth (Figure 8.2).

The woody trunk of a tree is made up of nonphotosynthetic tissues and, therefore, represents an investment and continuing expenditure of organic compounds produced by photosynthetic tissues. Thus, in extreme environments where the photosynthetic output must be conserved, the tree growth-form is not a viable possibility. In most situations it is thought that high winds and short growing seasons are the key environmental factors that act in combination to preclude the growth of trees in the tundra.

There are several observations that indicate the importance of wind. One of these is the interfingering of the Boreal forest and the Arctic tundra communities, with sheltered valleys supporting Boreal forest vegetation and windswept ridges having tundra. Another observation is that as treeline is approached from the south, the trees often have a shorter and wider appearance, even though they may be quite old. Drifting snow covers the lower branches in most winters, and they are in good condition; however, the upper limbs are often bare of foliage, especially on the side facing the prevailing winds. The exposed foliage is subject to ice abrasion and desiccation. Heat loss from summer winds is also important because it lowers metabolic activity. This combines with the short growing season to prohibit the maturation of new leaves. Immature leaves are especially subject to frost damage, ice abrasion, and desiccation if exposed during the winter. Thus, the short, spreading shape of the trees is adaptive, although it is not clear whether it is genetic or environmentally induced.

Another adaptation of the treeline species, at least the black and white spruce, is that much of the reproduction at timberline is by layering (branches rooting to form new individuals). This is advantageous, since in most years the environment is not conducive to regeneration by seed.

Human Impact. As with the tundra formation, the low productivity and relative isolation of much of the Boreal coniferous forest have limited its use by both aboriginal and modern man. The southern portion, especially in southeast Canada and northcentral and northeast United States, has been and is being logged. Most of the wood is used as pulp for paper products. Removal of mineral and energy resources has thus far been restricted to a few areas, but it is growing and will continue to do so as the earth's more accessible resources are depleted.

MONTANE CONIFEROUS FOREST—APPALACHIAN MOUNTAINS

Distribution

This coniferous forest forms a narrow band along the higher mountains and ridges of the Appalachian Mountains. It is quite similar to the Boreal forest. The two intermix in southeast Canada and northeast United States. In the mountains of Vermont and New Hampshire, the Appalachian coniferous forest occurs between the elevations of 800 and 1200 meters. Further south, it is found only at higher elevations: above 1100 meters in southeastern New York state and 1600 meters in the Great Smokey Mountains of Tennessee and North Carolina. Throughout most of this range, there is deciduous forest vegetation downslope, and the coniferous forest appears as a dark, high elevation band (Figure 8.7). Only the mountains of the northeastern United States have an Alpine tundra above the forest. The Allegheny Mountains of Pennsylvania, West Virginia, and Virginia are in the middle of the range of

FIGURE 8.7 The Appalachian coniferous forest appears as a dark band along the ridges and peaks as seen from North Carolina's Mt. Mitchell, the highest point in the eastern United States.

the Appalachian coniferous forest, but their low elevations preclude extensive stands of conifers. However, during the cooler temperatures of the Pleistocene, well-developed coniferous forest vegetation was continuous across this region.

Environment

There are many similarities as well as differences between the Boreal and the Appalachian coniferous forest environments. The climates of both forests are relatively cool, with precipitation exceeding potential evaporation. Temperatures are warmer in the Appalachian coniferous forest where they average around −5°C for the coldest month and 15°C for the warmest month. There is little difference in average figures between the northern and southern portions of the Appalachian coniferous forest, but there are greater temperature extremes in the north. Annual precipitation throughout the Appalachian coniferous forest is much higher than in the Boreal forest region. Average values increase from 150 centimeters in the north to 225 centimeters in the south, but there is less snowfall in the south. Summer precipitation in the form of fog-drip may be common throughout the mountains; there is one report of it increasing precipitation by two-thirds at a northern location.

Podzol soils have been found over the range of the forest. These are generally shallow, with bedrock less than one meter beneath the surface. The organic layer tends to be thicker in the north. Relief is highly variable and results in good drainage. The mountains do not have the numerous lakes and bogs that characterize the Boreal forest region.

Vegetation

Structure and Composition.   Since the Appalachian coniferous forest is also a spruce-fir type, it is often considered to be a lobe of the Boreal forest (Figure 8.8). Balsam fir, a Boreal forest dominant, is the most important fir in the northern part of the mountains. In the south, there is Fraser fir *(Abies fraseri)*, a closely related species that is presumed to have the same ecological role (such species are known as **ecological equivalents**). Fraser fir is thought to have evolved following post-Pleistocene isolation. The dominant spruce is

FIGURE 8.8   The interior of an Appalachian coniferous forest stand in mid March in Great Smoky Mountains National Park, North Carolina.

red spruce *(Picea rubens)*. It occurs throughout the mountains and extends north into part of the Boreal forest.

Birches are also present in the Appalachian coniferous forest; the most common are paper birch in the north and yellow birch *(Betula lutea)* throughout the range. They are usually considered to be seral species. Evergreen heath shrubs such as various azaleas and rhododendrons (both *Rhododendron* spp.) are common and may form treeless "heath balds" on exposed slopes. Herbaceous species are not as widespread as in the deciduous forest downslope. The surface layer of mosses and lichens is well developed, especially on tree limbs, trunks, and fallen logs. Many understory species are either the same as or closely related to Boreal forest species, or are common in the adjacent deciduous forest.

Adaptations.   There are a few adaptations to add to those discussed for the Boreal forest. One is the fanlike array of the needles of coniferous trees. As fog blows through the foliage, water droplets coalesce on the needles and drop, increasing soil moisture in the tree's root zone. A second additional adaptation is the branching of conifers. Their pointed crown and flexible limbs prevent the accumulation of snow, a very important factor for evergreens. Instead of building up on limbs and causing breakage, the snow slides off the flexible branches from limb to limb to the ground (photograph *b*, page 97). The fact that conifers also have pointed tops in low snowfall areas indicates that snow is not necessarily the key environmental factor in selection for this shape. A general but unexplained observation is that coniferous trees are more narrowly cone shaped in extreme environments than in moderate environments (assuming light conditions are equal). Lastly, many of the dominant heath shrubs have leaves that droop and curl in the winter; this is thought to reduce transpiration.

Human Impact.   Since the eastern United States is densely populated, the Appalachian coniferous forest has been greatly modified by human activities. Most of the forest has been logged, especially in the north where stands are more extensive. A second major impact has resulted from the great increase in the construction of vacation homes and resort developments in the last two decades.

MONTANE CONIFEROUS FOREST—ROCKY MOUNTAINS

Distribution

The Rocky Mountain coniferous forest extends from northern Alberta to New Mexico and from western South Dakota to central Utah and the eastern edge of the Cascade Mountains (Figure 8.1). It is not continuous throughout this range, and outliers occur—for example, on the slopes of some of the inter-mittent mountains of the Great Basin (Figure 1.2). In the north the Rocky Mountain forest blends into the Boreal, Northwest Coastal, and Sierra Nevada and Cascade Mountain coniferous forests. In the south, much of the forest borders woodland vegetation of the Pinyon-Juniper type (page 205). The montane regions of Mexico have vegetation that resembles that of the Rocky Mountains (page 238).

The vegetation of the Rocky Mountains is divisible into a series of altitudinal bands known as the alpine, subalpine, upper montane, lower montane, and foothill zones. Tundra vegetation occurs in the alpine zone, and Pinyon-Juniper woodland dominates the foothills. The coniferous forest covers the subalpine, upper montane, and lower montane zones, each of which has an elevational range of 650 meters or more. This zonation provides a framework for the study of mountain vegetation, but it should be recog-nized that the bands are at least somewhat subjective.

The elevational position of the bands is quite variable. Most importantly, they decrease in elevation with higher latitude. In fact, the lower zones are lost north of the central United States. Only the alpine, subalpine, and perhaps upper montane bands can be recognized near the northern limit of the forest. The zones are also affected by topographic relief; in general, they extend downslope on east-facing and north-facing slopes and in narrow ravines and valleys subject to cold air drainage.

Environment

The Rocky Mountain climate is the most continental of the western mountain chains (Figure 8.9). It is also highly variable. There is a northward decrease in temperature that explains the lower elevation of vegetation zones. The prevailing Westerlies produce significantly higher precipitation in the west-ern portion of the Rocky Mountains than on the eastern slope. Elevational differences in climate include decreases in temperature with increasing elevation. An average drop of 1.7°C per 300 meters has been reported. Such temperature differences change the growing seasons by two to three weeks per 300 meters in the central Rocky Mountains. Precipitation also varies elevationally. In many of the higher mountains it is at a maximum somewhat above midslope. Average annual values generally range from 40 to 150 centimeters. Winds are at a maximum at timberline, the upper elevational limit of trees, especially in the winter. They are also very strong in the foothills.

The soils of the Rocky Mountain coniferous forest are quite varied, but in general they are young and have a high rock content. They are usually deepest in the lower montane zone, where soil-formation processes are not greatly slowed by the low temperatures and low precipitation characteristic

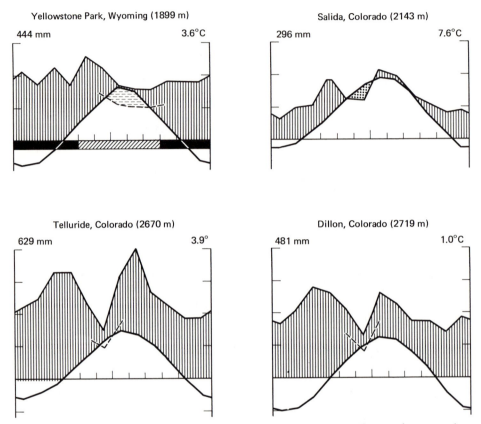

FIGURE 8.9 Climate diagrams for the Rocky Mountain coniferous forest region: Yellowstone Park is the furthest north of these locations; Salida is the furthest southeast; Telluride has a latitude close to that of Salida, but is further west and at a higher elevation; Dillon differs from Telluride in being nearer the drier, east slope of the mountains. The dotted lines indicate periods of water deficiency. *Source.* Redrawn from Walter, H. and H. Lieth. 1967. Klimadiagramm-weltatlas. Fischer, Jena, East Germany.

of other elevations. Organic matter, nitrogen levels, and acidity usually increase with elevation.

Topographic relief is another important environmental factor. Different slope exposures at the same elevation in the same region can support very different types of vegetation. In addition, narrow ravines and valleys—sites of cold air drainage in the evening—may have inverted vegetation zonation. Upper elevation species are often in the drainage, and species typical of lower elevations are higher on the slopes. Relief also interacts with heavy snowfall, since steep slopes may have avalanches periodically.

Lastly, fire has long been a reoccurring, natural part of the forest environment in the Rocky Mountains.

Vegetation

Composition and Structure.   The subalpine zone of the Rocky Mountains is very similar to the previous two coniferous forests in that the mature stands are largely spruce-fir communities. Engelmann spruce *(Picea engelmanii)* and subalpine fir *(Abies lasiocarpa)* are the climax dominants throughout most of the range, and white spruce can be important in northern stands (Figure 8.10). Different species are frequently dominant at timberline; bristlecone pine *(Pinus aristata)*, the species with the world's oldest living organisms, occurs at timberline in the south (Figure 1.2). In the upper montane zone, the chief dominant of mature stands is douglas fir *(Pseudotsuga menziesii)*, a species of wide distribution.

In many subalpine and upper montane stands, the above species are neither dominant nor even common. Instead, lodgepole pine *(Pinus contorta)* and quaking aspen are frequently most important. Both of these are usually considered seral species. Lodgepole pine is more typical of drier sites and is not found in the south; quaking aspen occurs on mesic sites throughout the range of the forest. Many stands of these two species are even aged (i.e.,

FIGURE 8.10   The subalpine and alpine zones near Berthoud Pass in northcentral Colorado in mid June. Note the decrease in height of subalpine fir toward timberline.

most individuals are the same age). This is indicative of an origin following disturbance. Fires and logging were and are widespread but, locally, avalanches and insect outbreaks also have been important. The end result is a mosaic of seral and mature stands in both the subalpine and upper montane zones (Figure 8.11).

At lower elevations, douglas fir forms a broad transition between the two montane zones. However, open stands of ponderosa pine *(Pinus ponderosa)* with a dense herbaceous layer dominated by grasses are characteristic of the lower montane zone (Figure 8.12).

As indicated by the presence of douglas fir in both montane belts, zonation is a simplified picture of mountain vegetation. Species distributions are actually very complex and depend on the interaction of species tolerance limits and microenvironments that are produced by variations in relief and latitude. Species tend to be present primarily on warm slopes or in seral stands at their upper or northern extremes and in mesic sites at their lower or southern extremes. This indicates that the upper elevational limits of montane species are usually determined by low temperatures and that lower limits depend on soil factors, especially moisture. Locally, other climate, soil, and biological factors may be more important.

FIGURE 8.11 The mosaic of dark-colored mature stands of conifers and light-colored seral stands of quaking aspen in the upper montane and subalpine zones of the Rocky Mountains near Aspen, Colorado. A montane meadow is in the foreground.

FIGURE 8.12 A stand of ponderosa pine in the lower montane zone of the Rocky Mountains near Flagstaff, Arizona. Note the open nature of the forest.

Adaptations. The adaptations are primarily the same as those covered in earlier portions of this chapter. For example, the shallow root system of most tree species was mentioned as an important factor where soils are poorly developed. Shallow roots are also adaptive to low soil moisture. Species of lower elevations reportedly are less sensitive to drought and have a rapid rate of root elongation.

The quaking aspen's root sprouting was previously described because of its ability to rapidly revegetate disturbed areas (page 106). Data of L. L. Loope and G. E. Gruell show that quaking aspen in Wyoming may produce 12,000 to 20,000 stems per hectare. It is thought that fire produces a physiological stimulus to root sprouting. The sprouts grow rapidly, and competition quickly reduces their density.

In many regions of the Rocky Mountains, lodgepole pine resembles the jack pine of the Boreal forest in having serotinous cones. Temperatures of 45°C are required to open them. Even in regions where lodgepole pine is not serotinous, its reproductive capacity is so great that it is still a common seral species. Other adaptations to fire include the thick, somewhat fire-resistant bark of such species as douglas fir and ponderosa pine. Of the two, douglas fir is more frequently killed by fire, since it grows in dense stands where greater amounts of combustible material produce hotter and longer lasting fires than in naturally open ponderosa pine forests.

Montane herbs have many of the adaptations discussed for the Alpine tundra (Chapter 7). In addition, some subalpine wet meadow species such as broad-leaved lungwort *(Mertensia ciliata)* have hollow stems. Research by W. D. Billings and P. J. Godfrey has shown that inside these stems daytime temperatures may be 20°C warmer and CO_2 concentrations 2 to 50 times greater than outside. The CO_2 is produced by respiration of stem and root cells and is recycled by photosynthesis carried on inside the stem. Such internal photosynthesis is significant throughout the growing season but is especially important in early spring, when respiration is very high and before leaves have formed and external air temperatures have warmed. Hollow stems are also advantageous in that they are a structural method of conserving carbohydrates.

Timberline.   As with the northern edge of the Boreal forest, subalpine trees in all major North American mountain chains become shorter toward timber-line (Figure 8.10). At this upper elevational limit, the tree growth-form is shrubby; trees have lateral but little vertical growth (Figure 8.13). This distinctive growth-form is known as krummholz. Apparently, as at the treeline in the Boreal forest, the growing season is so short and strong winds cause such heat loss that new foliage is not able to mature in one season. As a result, while branches covered by snow may survive the winters (hence the krummholz growth-form is adaptive), exposed foliage suffers frost damage, snow and ice abrasion, and desiccation. This effect would be greatest at timberline, since winds are reportedly stronger there than at any other forested elevation. The krummholz growth-form may be genetically determined.

Human Impact.   Prior to European settlement, fires periodically burned sections of the Rocky Mountains. This produced and maintained diverse ecosystems with a variety of habitats for animal species, some of which are dependent on seral stands. Many of the fires were surface fires, ones which burn along the ground and do not reach the canopy. These fires maintained open, parklike conditions with low levels of combustible forest materials (dead branches, foliage, etc.).

For the last several decades, fire suppression has effectively reduced the

FIGURE 8.13 The krummholz growth-form of Engelmann spruce at timberline (3250 m elevation) in mid June in Rocky Mountain National Park, Colorado. The trees in the foreground are less than a meter tall, and those further downslope (left) are several meters.

frequency and spread of fires. Without periodic surface fires, most of today's stands are denser and have increased fuel levels. As a result of fire protection, the probability of crown fires, fires which burn into the canopy, has greatly increased. Crown fires are very destructive, especially today when many unlogged areas are now so mature that they lack seral species to quickly and effectively revegetate an area following fire. Also, since increased canopy densities have resulted in reduced ground cover beneath them, the soil is more susceptible to accelerated erosion following fire.

Other problems have resulted from the logging of much of the forest. In spite of attempts to improve logging methods, there is widespread concern for present-day practices. Some of the most difficult problems include cutting and removing trees from remote stands in mountainous terrain, reducing erosion following loss of the tree canopy and disturbance of the soil, eliminating the fire hazard of the dry slash (dead tops and limbs not used for lumber), revegetating cut-over areas, intermixing a variety of land uses in a single area, and achieving sustained yield so that there will be a continual supply of timber in the future.

Other significant impacts are the result of grazing, mining, and resort development. Sheep and cattle grazing have been especially common in meadows and in the ponderosa pine forests, where the herbaceous layer is dense (Figures 8.11 and 8.12). In some areas this grazing has caused accelerated erosion. In other areas grazing and logging have led to the establishment of the dense scrub vegetation of the Oak-Mountain Mahogany type (page 211) in place of the original ponderosa pine forest. Mining is destructive where it has occurred. The abandoned silver mine areas of Colorado provide clear evidence of this. The impact of resort developments is increasing as resorts become more widespread. The erosion of ski slopes with the spring snow melt, versions of suburban sprawl, and effective treatment of sewage are problems common to many resorts in the Rocky Mountains and elsewhere.

MONTANE CONIFEROUS FOREST—SIERRA NEVADA AND CASCADE MOUNTAINS

Distribution

The coniferous forest of the Sierra Nevada and Cascade Mountains has a much narrower distribution than that of the Rocky Mountains (Figure 8.1). Its northern limit is in British Columbia, where it meets the Northwest Coastal and Rocky Mountain coniferous forests. In southern California, it is bordered by other formations at lower elevations.

An elevational vegetation banding can be described but, as a result of the moderating influence of the Pacific Ocean, the zones are lower in elevation than in the Rocky Mountains. The Alpine tundra zone occurs above about 2000 meters in the north and 3500 meters in the south. Below it there are three coniferous forest zones: subalpine, upper montane, and lower montane. Further downslope is the foothill zone; it has woodland and shrubland types to the west and desert vegetation to the east. The zones are compressed and often indistinct on the drier, steep east slope of the mountains.

Environment

The environment of this forest region is dominated by the Mediterranean climate found in most of far western North America. It is characterized by wet, cool winters and dry, warm summers (Figure 8.14). The precipitation gradient ranges from 25 to 35 centimeters at the base of the west slope to a maximum of over 125 centimeters in the upper montane or subalpine zone. In the central Sierra Nevada, roughly 80 to 85 percent of the precipitation falls in the winter. This percentage drops to approximately 60 to 65 percent further north. Except for the foothill zone, nearly all winter precipitation falls as snow; there may be enormous amounts (10 to 20 meters annually) in the subalpine zone. Throughout the mountains, little precipitation occurs dur-

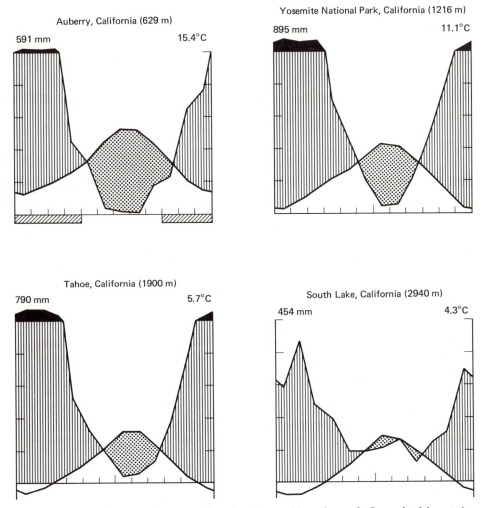

FIGURE 8.14   Climate diagrams for the Sierra Nevada and Cascade Mountains coniferous forest region: these locations are along an elevational gradient from the west base to the east slope of the central Sierra Nevada. *Source*. Redrawn from Walter, H. and H. Lieth. 1967. Klimadiagramm-weltatlas. Fischer, Jena, East Germany.

ing the summer, and dry conditions may be a limiting factor for growth. Summer drought is somewhat less severe in the north.

The soils of this forest type are quite varied but are generally young (i.e., thin with a high rock content). Dry summers and cool temperatures slow the rate of soil formation. Relief is highly variable and influences the elevations of the vegetation zones. Fire has long been an important environmental factor in this region, as indicated by basal fire scars on most old trees. Lightning strike fires are very common, especially during the dry summers. Additional fires were formerly set by Indians to aid in hunting and gathering of food supplies. The great majority of all presettlement fires were surface fires. Depending on location, intervals of only 3 to 15 years between fires were typical in the past.

Vegetation

Composition and Structure.   Unlike the upper zone of the two other montane coniferous forests, the subalpine zone of the Sierra Nevada and Cascade Mountains is not a spruce-fir forest. Instead, most mature stands are dominated by mountain hemlock *(Tsuga mertensiana)* and lodgepole pine. The latter is a different variety from the one in the Rocky Mountains. Other species may be common in various sites, especially at the timberline where low species diversity allows less effective competitors to dominate. One of the more widespread timberline species is whitebark pine *(Pinus albicaulis)*. Red fir *(Abies magnifica)* may occur in subalpine stands, but it typically dominates large stands that are intermediate between the subalpine and upper montane zones (photograph *c,* page 98).

The true upper montane zone supports a diverse forest. The two primary dominants are douglas fir and white fir *(Abies concolor)*; the former is absent in the southern Sierra Nevada. Sugar pine *(Pinus lambertiana)* and incense cedar *(Calocedrus decurrens)* are widespread in certain portions of the zone. Giant sequoia *(Sequoiadendron giganteum)* is dominant in scattered groves in the central and southern Sierra Nevada. Individuals of this species are the largest living organisms on earth (Figure 8.15).

The lower montane zone is similar to that of the Rocky Mountains. Ponderosa pine forms an open forest; shrubs, many from the foothill zone, and grasses compose the understory. Similar forests of Jeffrey pine *(Pinus jeffreyi)* often occur at somewhat higher elevations. Both of these yellow pines extend into the upper montane zone, and incense cedar frequently is found in the lower montane zone.

On the drier, steep east side of the mountains, any recognizable zones are generally less diverse and often contain cold desert species such as sagebrush *(Artemisia tridentata)*.

Adaptations.   The thick cuticle, strong epidermis, and sunken stomates of evergreen conifer leaves are highly adaptive to the dry summers of this forest region. In addition, adaptations to fire are critical. Several montane shrub species have seeds whose germination is stimulated by exposure to fire. Also, most conifers have seeds that germinate and grow best on ashy,

FIGURE 8.15 The "General Sherman" giant sequoia, the world's largest organism. It is located in the upper montane zone of Sequoia National Park, California. The area in front of the tree has been artificially cleared. Other tree species in the photograph are white fir and sugar pine.

mineral soils—the type produced when fires burn the organic matter on the soil surface. Thick bark is common but is best developed on giant sequoia, where it may be 30 to 50 centimeters in thickness and nearly asbestoslike in its resistance to burning. Large fire scars at the bases of these trees provide evidence of frequent fires throughout their life spans of 2000 to 3000 years. The longevity of giant sequoias is aided by virtual immunity to disease and insect damage.

Human Impact. Human influence on this forest was substantial long before European settlement. As mentioned before, the Indians of the region regularly set fires. These and lightning strike fires produced an open, parklike aspect in much of the forest, especially in the Sierra Nevada. As a result of extensive fire suppression during this century, the forests have developed dense tangles of undergrowth and high proportions of shade tolerant species (Figure 8.16). Today, the majority of fires are not the surface type; because of the high amount of combustible material extending from the floor to the canopy, they quickly convert into destructive crown fires. These are very difficult to suppress and, depending on intensity, can lead to forest replacement by shrub vegetation of the Chaparral type (page 213).

Recently, as in other areas where fires have been suppressed, attempts have been made to restore fire to ecosystems. Prescribed burning—the setting of small controlled fires under certain weather conditions—has been used to reduce forest fuel levels in some locations. In a few other carefully deliniated areas, where the danger of a widespread crown fire is not great, lightning fires are let go until they burn themselves out. Today, the objective of fire management in some areas is to return to the point where fire is a natural part of the environment.

Most of the forest is publicly owned and included in National Forests. These are managed under the "multiple-use" concept whereby the forests are intended to be used for a variety of purposes, including recreation, grazing, and timber production. However, the last activity has tended to predominate. Ponderosa pine and douglas fir are very important commercial timber species. Livestock grazing is especially common in meadows and at lower elevations where the herbaceous layer is well developed. Grazing was more intense in the past and sometimes resulted in significant erosion problems and increased forest densities (through exposure of the mineral soil).

Past mining activities also have had a persisting impact on the forest. Mining was especially common in the lower elevations of the California region beginning with the gold rush in 1849. The revegetation of areas disturbed by surface mining may take many decades.

More recently, the growing recreational use of the mountains has had adverse effects. In areas such as Lake Tahoe, resorts have been expanded with little regard for land-use principles. Public campgrounds frequently fill to capacity and have population densities greater than the most crowded cities. The sport of backpacking has become so popular that trail erosion, firewood gathering, and unsanitary conditions have become problems in many areas.

Lastly, air pollution is becoming an increasingly important factor adversely affecting this forest. Pollution levels are greatest in the central and southern portions of the Sierra Nevada, near cities such as Los Angeles, Bakersfield, and Fresno. The lower-elevation vegetation is most severely affected. Species vary in their susceptibility to air pollution; ponderosa pine is especially subject to damage and eventually to death. Studies by P. L. Miller in southern California indicate that in some areas shrublands will likely replace forests where tree damage is severe.

FIGURE 8.16 Comparison photographs of the same view in the upper montane zone of Sequoia National Park, California. The top photograph was taken in 1912; the other was taken in 1968 after several decades of effective fire suppression. *Sources.* (top) The Sierra Club collection of the Bancroft Library, University of California, Berkeley; (bottom) the author.

NORTHWEST COASTAL CONIFEROUS FOREST

Distribution

This coniferous forest forms a narrow belt along the western coastline of the continent from Alaska to central California; it seldom extends more than 175 kilometers inland (Figure 8.1). In elevation, the forest ranges from sea level to several thousand meters. Along much of its length it parallels, if not borders, the coniferous forest of the Sierra Nevada and Cascade Mountains. In the north it meets both the Rocky Mountain and Boreal coniferous forests. In the south it borders several woodland and shrubland vegetation types.

Environment

The climate of the Northwest Coastal forest is greatly influenced by the moderating effect of the Pacific Ocean (Figure 8.17). Temperatures are consistently mild. Even in the far north, temperatures below $-15°C$ are uncommon along the coast. Frosts are rare in the southern portion of the range.

Precipitation is so great that the forest is sometimes referred to as a temperate rain forest. Values may vary from 65 to 400 centimeters or more; the lower amounts occur inland. In the south, over 75 percent of the precipitation falls during the winter; in the north, seasonal variation is insignificant. Nearly all of the precipitation in the south falls as rain but, in the north, especially in the coastal mountains, annual snowfall may total many meters.

Although summer months in the south have little precipitation, atmospheric humidity remains high. In part, this is a result of the moderation of temperatures by the ocean and by fogs. Fogs may occur many kilometers inland. They are important in summer dry climates, since they reduce evapotranspiration by decreasing the penetration of sunshine and by maintaining lower daytime temperatures. They also may add to soil moisture through fog-drip, but this is probably not as important a factor as reduced evapotranspiration. The significance of fog decreases to the north, as summer drought is reduced and is finally nonexistent north of central British Columbia.

The soils of the Northwest Coastal forest region are quite variable, but Podzols are common, since they form under relatively cool temperatures and high amounts of precipitation. Topographic relief is also variable, ranging from the flats of river flood plains to the steep slopes of the coastal mountains, which occur along the entire length of the forest. Many of the coastal mountains are low in elevation, but higher peaks with Alpine tundra occur from northwest Washington to Alaska. Fire is important throughout most of this forest type.

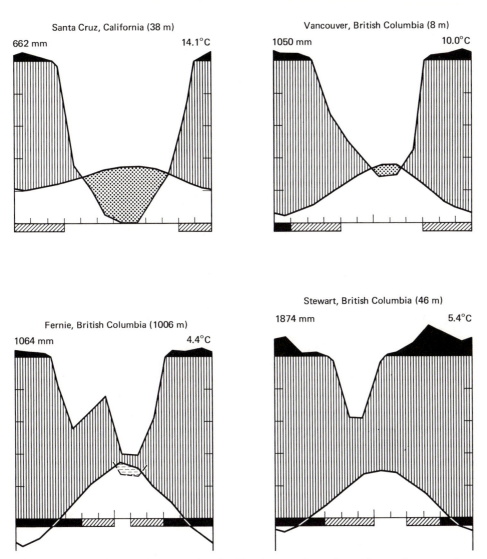

FIGURE 8.17 Climate diagrams for the Northwest Coastal coniferous forest region: Santa Cruz, Vancouver, and Stewart are coastal locations along a south to north gradient; Fernie is east of Vancouver in a region where this forest type extends inland. *Source.* Redrawn from Walter, H. and H. Lieth. 1967. Klimadiagramm-weltatlas. Fischer, Jena, East Germany.

FIGURE 8.18   The interior of a Sitka spruce stand of the Northwest Coastal coniferous forest in southern Oregon.

Vegetation

Composition and Structure.   The Northwest Coastal forest has the tallest stands in the world; they are also very dense (Figure 8.18). Canopy trees are often over 75 meters in height and 2 meters in diameter. A subcanopy layer dominated by angiosperms is usually present, which makes the structure more complex than that of the typical coniferous forest. The shrub layer is dense, but the herbaceous layer is not well developed. Mosses are common.

Although the appearance of the forest is quite consistent throughout its range, species composition varies. Maximum development of the forest occurs from central British Columbia to southern Oregon. Part of this region is thought to have served as a refugium where species survived during Pleistocene glaciation and then migrated northward following the retreat of

FIGURE 8.19   The interior of a redwood stand of the Northwest Coastal coniferous forest in northern California.

the glaciers. The dominant climax species are western hemlock *(Tsuga hetero-phylla)*, Sitka spruce *(Picea sitchensis)*, and western arbor vitae *(Thuja plicata)*, with some firs also important. However, the most common species is douglas fir, which is seral and forms pure, even-aged stands, reflecting an origin following fire.

The northern portion of the forest is dominated by western hemlock, with Sitka spruce along the coastline and mountain hemlock where the flora mixes with that of the Sierra Nevada and Cascade Mountains forest. White spruce, primarily a Boreal forest species, is found in the extreme north. The southern portion of the Northwest Coastal forest is dominated by redwood *(Sequoia sempervirens*; Figure 8.19). Individuals of this species are the tallest organisms on earth; the current record holder is 117 meters in height. Coastal fog is apparently important in the distribution of redwoods, since the species is found only in areas where fog occurs frequently (photograph *d,* page 98, shows remnants of the evening fog). Stands with redwood as the only dominant occur in coastal valleys and river flats. Douglas fir and other species are more important on slopes.

In the middle of the range of this forest type, several of the dominant species extend eastward along the United States–Canada border. Apparently, this is the result of the maritime climate extending inland along a pathway of coastal air masses. The major vegetational gradient in this region is the increasing importance of douglas fir toward the Rocky Mountains.

Adaptations. Perhaps the most interesting species of the Northwest Coastal forest is the redwood. It is widely used as a commercial timber species because of its great size and resistance to decay and insect damage. Redwoods can root sprout; this is especially advantageous for the revegetation of land after it has been logged or otherwise disturbed. The tree's bark is thick and resistant to fire. Douglas fir is also fire resistant in this manner, but the hemlocks and arbor vitae that are the climax species in much of the forest are not. Douglas fir is also a prolific producer of seeds that germinate best on a mineral soil. Thus, this fire-dependent species is well adapted to its seral role. The seeds of some climax species are capable of germinating and growing on fallen logs, stumps, and so on, as in an old age forest.

Human Impact. A major impact has been exhibited as most of the old growth stands, especially in the central and southern portions of the forest, have been logged. The cutting of douglas fir forests in Washington and Oregon alone accounts for over one-third of the annual lumber production of the United States. This makes douglas fir the most important commercial timber tree in North America. Most of the logged areas of the Northwest Coastal forest were clear-cut, either entirely or in a checkerboard pattern (Figure 8.20). The short-run commercial advantage of clear-cutting is obvious, but the consequent exposure of the soil can result in erosion and revegetation problems.

Today, less than 5 percent of the mature stands of redwood are uncut; only one-half of these have been preserved in state and national parks. Logging adjacent to these parks has resulted in greater water runoff and flooding of the redwoods along their river-flat habitat. Natural flooding in the past provided a source of nutrients for these rapidly growing trees, but the damage caused by accelerated flooding may outweigh this benefit.

Another human impact has been the changed role of fire in these forests. Fire suppression has led to increased forest densities and higher fuel loads. The danger of fire is also increased by the dead, dried slash left after logging. Prescribed burning is sometimes used to reduce fire hazards.

FIGURE 8.20 The logging of forest stands may involve clear-cutting, block clear-cutting, or selective cutting. The top photograph shows an area in the southern portion of the Northwest Coastal coniferous forest region that had been clear-cut a few years before. Note the great exposure of soil. The bottom photograph shows a block clear-cutting in a Boreal coniferous forest stand in northern Maine. Selective cutting involves the periodic cutting of mature trees on an individual basis, leaving all others.

129

SUGGESTED READINGS FOR FURTHER STUDY

Billings, W. D. and P. J. Godfrey. 1967. Photosynthetic utilization of internal carbon dioxide by hollow-stemmed plants. Science 158:121–123.

Buell, M. F. and W. A. Niering. 1957. Fir-spruce-birch forest in northern Minnesota. Ecology 38:602–610.

Cooper, C. F. 1960. Changes in vegetation, structure, and growth of southwestern pine forests since white settlement. Ecological Monographs 30:129–164.

Dansereau, P. and F. Segadas-Vianna. 1952. Ecological study of the peat bogs of eastern North America. Canadian Journal of Botany 30:490–520.

Daubenmire, R. F. 1943. Vegetational zonation in the Rocky Mountains. Botanical Review 9:325–393.

———. 1952. Forest vegetation of northern Idaho and adjacent Washington, and its bearing on concepts of vegetation classification. Ecological Monographs 22:301–330.

Davis, R. B. 1966. Spruce-fir forests of the coast of Maine. Ecological Monographs 36:79–94.

Fonda, R. W. 1974. Forest succession in relation to river terrace development in Olympic National Park, Washington. Ecology 55:927–942.

Fonda, R. W. and L. C. Bliss. 1969. Forest vegetation of the montane and subalpine zones, Olympic Mountains, Washington. Ecological Monographs 39:271–301.

Franklin, J. F. and C. T. Dyrness. 1973. Natural vegetation of Oregon and Washington. United States Department of Agriculture. Forest Service Technical Report PNW–8. 417 p.

Habeck, J. R. and R. W. Mutch. 1973. Fire-dependent forests in the northern Rocky Mountains. Quaternary Research 3:408–424.

Hartesveldt, R. J., H. T. Harvey, H. S. Shellhammer, and R. E. Stecker. 1975. The giant sequoia of the Sierra Nevada. United States Department of the Interior. National Park Service Publication Number 120, Washington, D.C. 180 p.

Kilgore, B. M. 1973. The ecological role of fire in Sierran conifer forests: Its application to National Park management. Quaternary Research 3:496–513.

LaRoi, G. H. 1967. Ecological studies in the boreal spruce-fir forests of the North American taiga. I. Analysis of the vascular flora. Ecological Monographs 37:229–253.

Larsen, J. A. 1974. Ecology of the northern continental forest boundary, p. 341–369. *In* J. D. Ives and R. G. Barry (eds.). Arctic and alpine environments. Methuen, London.

Loope, L. L. and G. E. Gruell. 1973. The ecological role of fire in the Jackson Hole area, northwestern Wyoming. Quaternary Research 3:425–443.

Marr, J. W. 1948. Ecology of the forest-tundra ecotone on the east coast of Hudson Bay. Ecological Monographs 18:117–144.

Maycock, P. F. and J. T. Curtis. 1960. The phytosociology of boreal conifer-hardwood forests of the Great Lakes region. Ecological Monographs 30:1–35.

McIntosh, R. P. and R. T. Hurley. 1964. The spruce-fir forests of the Catskill Mountains. Ecology 45:314–326.

Miller, P. L. 1973. Oxidant-induced community change in a mixed conifer forest, p.

101–117. *In* J. A. Naegele (ed.). Air pollution damage to vegetation. Advances in Chemistry Series 122. American Chemical Society, Washington, D.C.

Oosting, H. J. and W. D. Billings. 1951. A comparison of virgin spruce-fir forest in the northern and southern Appalachian system. Ecology 32:84–103.

Rowe, J. W. and G. W. Scotter. 1973. Fire in the boreal forest. Quaternary Research 3:444–464.

Rundel, P. W., D. J. Parsons, and D. T. Gordon. 1977. Montane and subalpine vegetation of the Sierra Nevada and Cascade Ranges, p. 559–599. *In* M. G. Barbour and J. Major (eds.). Terrestrial vegetation of California. Wiley, New York.

Vankat, J. L. and J. Major. 1978. Vegetation changes in Sequoia National Park, California. Journal of Biogeography 5:377–402.

Wardle, P. 1974. Alpine timberlines, p. 371–402. *In* J. D. Ives and R. G. Barry (eds.). Arctic and alpine environments. Methuen, London.

Waring, R. H. and J. Major. 1964. Some vegetation of the California coastal redwood region in relation to gradients of moisture, nutrients, light, and temperature. Ecological Monographs 34:167–215.

Weaver, H. 1974. Effects of fire on temperate forests: Western United States, p. 279–319. *In* T. T. Kozlowski and C. E. Ahlgren (eds.). Fire and ecosystems. Academic Press, New York.

Whittaker, R. H. 1956. Vegetation of the Great Smoky Mountains. Ecological Monographs 26:1–80.

Zinke, P. J. 1977. The redwood forest and associated north coast forests, p. 679–698. *In* M. G. Barbour and J. Major (eds.). Terrestrial vegetation of California. Wiley, New York.

9

DECIDUOUS
FOREST VEGETATION

Mature stands of the deciduous forest of eastern North America usually have five vertical strata. There are two tree layers: a dense canopy and a more open subcanopy composed of immature trees and mature trees of species that are not as tall as those of the canopy (photograph *a,* page 133). Widespread examples of subcanopy species include flowering dogwood *(Cornus florida)* and redbud *(Cercis canadensis).* The lower layers are the shrub, herb, and surface strata. Perennial forbs dominate the herb layer (photograph *c,* page 133). The surface layer of mosses and lichens is present mainly on tree trunks and rocks.

One of the most striking aspects of the deciduous forest is its seasonal variation (Figure 9.1). During winter the leafless forest is open, and the plants are dormant. In spring new sets of leaves are formed and most plants flower. During summer the leaves give the forest a very dense appearance, and little light penetrates the canopy to the forest floor. In fall the leaves turn color and then drop.

DISTRIBUTION

Deciduous forests cover major land areas in only three parts of the world: central Europe, eastern Asia, and eastern North America.

In North America, the grassland formation bounds the deciduous forest from eastern Texas to western Minnesota (Figure 9.2). Low soil-moisture levels (and formerly prairie fires) are thought to limit tree growth in this transitional region. The boundary of the deciduous forest with the Boreal

(a) The interior of a mature forest in southwest Ohio.
(b) A second growth, sprout forest in southwest Pennsylvania.
(c) The herbaceous understory of a forest in northeast New York.
(d) A mature forest in the central Appalachian Mountains, Virginia.

FIGURE 9.1 A relatively young deciduous forest in southwest Ohio in (a) winter, (b) early spring, and (c) summer.

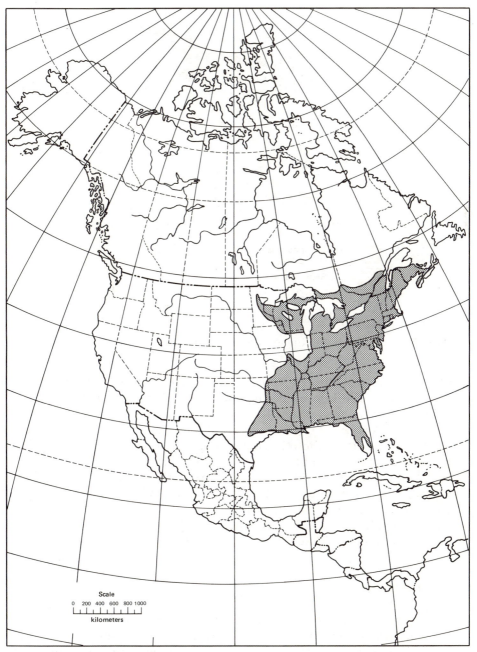

FIGURE 9.2 Distribution map of the deciduous forest.

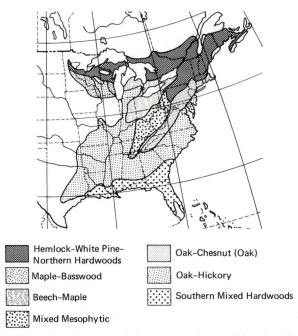

▨ Hemlock–White Pine–Northern Hardwoods	▦ Oak–Chesnut (Oak)
⣿ Maple–Basswood	⣿ Oak–Hickory
⣿ Beech–Maple	⣿ Southern Mixed Hardwoods
⣿ Mixed Mesophytic	

FIGURE 9.3 Distribution map of the associations of the deciduous forest.

forest to the north is obscure, but it extends from northern Minnesota to northern Maine. The key environmental factor in this boundary region is thought to be low temperatures that produce short growing seasons. To the east and south, except in Florida, the deciduous forest extends to the coastline of the continent. Southern Florida, with growing seasons nearly 365 days long, has vegetation resembling that of the tropics.

With such a wide distribution, the North American deciduous forest shows much variation in composition, and thus it can be divided into several relatively distinct associations: the Mixed Mesophytic, Beech-Maple, Maple-Basswood, Hemlock-White Pine-Northern Hardwoods, Oak-Chestnut, Oak-Hickory, and Southern Mixed Hardwoods Forests (Figure 9.3).

ENVIRONMENT

The large area covered by the deciduous forest and its division into several associations imply that the environment must have significant variation. However, the dominance of the deciduous tree growth-form and the widespread distribution of the oak *(Quercus)*, maple *(Acer)*, beech *(Fagus)*, and basswood *(Tilia)* genera indicate that there must also be basic environmental factors that are consistent throughout the range of the forest. Four such climatic factors are seasonal changes, frosts, absence of periodic droughts, and moderate to long growing seasons (Figure 9.4).

Precipitation generally decreases from east to west, with amounts of around 125 centimeters along the Atlantic Coast and 85 centimeters along the grassland-forest boundary. There also is a significant difference in precipitation from north to south, with values of 75 centimeters in the Great Lakes region and 150 centimeters along the Gulf Coast. Temperature differences result in much of this precipitation falling as snow in the north and nearly all as rain in the south. The length of the growing season also varies; it is around 120 days in the north and 250 days in the south.

In addition to regional differences in climate, stands show great vertical microclimatic variation. From the canopy in full leaf to the forest floor, there are great decreases in light intensity and wind velocity, a moderate decrease in air temperature, and a significant increase in atmospheric humidity. Also, a canopy in leaf intercepts a high percentage of the precipitation. Part of this reaches the ground by dripping or flowing down the bark of limbs and trunks, but much of it is lost in evaporation.

Two Great Soil types divide the North American deciduous forest in half. In the north there is Grey-brown Podzolic soil, and in the south there is Red-yellow Podzolic. In comparing the two, the Grey-brown Podzolic soils are more acid, have more organic matter, are more fertile, and are more similar to Podzols. The Red-yellow Podzolic soils are more similar to Laterites and, most importantly, are less fertile than the Grey-brown soils. Yellow soils are common in the Coastal Plain physiographic region; Red soils are predominant in the upland regions (Figure 4.9).

Topographic relief in the forest region is quite variable. Glaciation in the north produced a generally flat to rolling landscape. South of the limit of glaciation, the topography is more diverse. The Appalachian Mountains extend from the glaciated north into the unglaciated south. Fire has had a major impact on the vegetation in many areas, but especially along the grassland-forest boundary and in the South. Lastly, some biologic factors, such as plant diseases, have been extremely important, primarily in this century.

VEGETATION

Associations

Mixed Mesophytic Association. Mixed Mesophytic vegetation occurs on moist, well-drained sites in the central part of the forest. The range of the association includes a portion of eastern Tennessee, the eastern one-third of

Sault Ste. Marie, Michigan (220 m)
767 mm   4.0°C

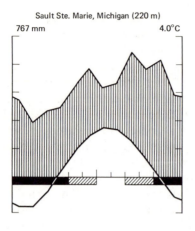

Dubuque, Iowa (196 m)
833 mm   8.3°C

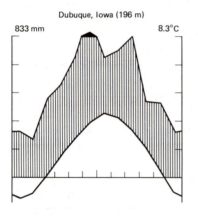

Dayton, Ohio (305 m)
893 mm   11.2°C

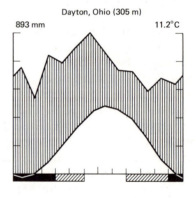

Charleston, West Virginia (289 m)
1143 mm   13.2°C

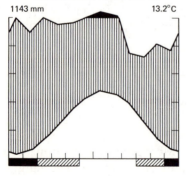

Knoxville, Tennessee (299 m)
1156 mm   15.2°C

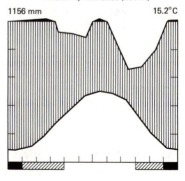

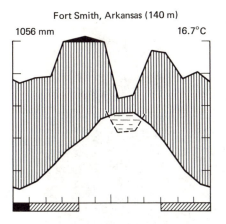

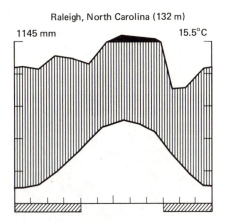

Tallahassee, Florida (20 m)

1439 mm   19.8°C

FIGURE 9.4 Climate diagrams for the deciduous forest region: Sault Ste. Marie is in the Hemlock-White Pine-Northern Hardwoods association; Dubuque is in the Maple-Basswood association; Dayton is in the Beech-Maple association; Charleston is in the Mixed Mesophytic association; Knoxville is in the Oak-Chestnut association; Fort Smith is in the western portion and Raleigh is in the eastern portion of the Oak-Hickory association; Tallahassee is in the Southern Mixed Hardwoods association. *Source.* Redrawn from Walter, H. and H. Lieth. 1967. Klimadiagramm-weltatlas. Fischer, Jena, East Germany.

FIGURE 9.5   An example of a cove hardwood, Mixed Mesophytic stand in the southern Appalachian Mountains of North Carolina.

Kentucky, southeastern Ohio, and much of West Virginia. It also extends into the narrow valleys (coves) of the southern Appalachian Mountains, where Oak-Chestnut vegetation predominates (Figure 9.5). To the west it blends with the Oak-Hickory association in a broad transition zone that is sometimes referred to as the Western Mesophytic forest region. This area, primarily in western Kentucky and Tennessee, is dominated by a mosaic of stands of Oak-Hickory and Mixed Mesophytic vegetation.

The Mixed Mesophytic association is the most diverse of the deciduous forest. Widespread dominants include American beech *(Fagus grandifolia)*, tuliptree *(Liriodendron tulipifera)*, several basswoods, sugar maple *(Acer saccharum)*, sweet buckeye *(Aesculus octandra)*, red oak *(Quercus rubra)*, white oak *(Quercus alba)*, and eastern hemlock *(Tsuga canadensis)*. In addition, there are 20 to 25 others that achieve dominance in particular stands—thus there is great stand diversity. Of all the species, the best indicators of this association are sweet buckeye and white basswood *(Tilia heterophylla)*. Species diversity is greatest in the Cumberland Mountains of eastern Tennessee and Kentucky.

Vegetation similar to the Mixed Mesophytic association is thought to have been important to the history of the deciduous forest formation. Comparison of the flora of this association with that of the eastern Arcto-Tertiary Geoflora shows a great resemblance. Approximately 80 percent of the genera of Mixed Mesophytic stands in the coves of the Great Smoky Mountains are of Tertiary age.

Some scientists have claimed that the Tertiary relationships and the presence of dominants of other associations are evidence that the Mixed Mesophytic is the oldest association of the deciduous forest and that it produced some of the other associations. However, it is now thought that the Mixed Mesophytic association did not remain intact during Pleistocene glaciation. Mixed Mesophytic species migrated southward, although the sites of refugia are not well known, and species returned northward following the retreat of the Wisconsin-stage glacier. These species migrations produced the presettlement composition of the Mixed Mesophytic and other associations.

Beech-Maple Association.   This association occurs only on glaciated land. Its region includes most of Ohio, western Indiana, and part of southern Michigan. The boundary with the Mixed Mesophytic association is relatively distinct, matching the southern extent of Wisconsin glaciation. Other boundaries are not as well defined.

Within its region, the Beech-Maple association is best developed on mesic, fairly well-drained sites (Figure 9.6). The two dominants are American beech and sugar maple; the former is usually somewhat more important in the canopy, and the latter is dominant in the understory. There are several

FIGURE 9.6   The interior of a mature Beech-Maple forest stand in southwest Ohio.

associated species, some of which are of greater importance in other habitats. For example, ashes (*Fraxinus* spp.) and American elm *(Ulmus americana)* in presettlement times dominated areas with high soil moisture, such as swampy river flood plains and other poorly drained areas (page 150).

The Beech-Maple association is relatively young, having been formed from species migrations following the retreat of the last continental glacier. Evidence of past climates is scattered throughout the region. For example, a few communities of Boreal forest species occur where microenvironmental conditions resemble a more northerly climate, such as in bogs. These communities are presumed to be relicts of the postglacial, northward migration of Boreal forest-type vegetation.

There also are relicts of the Hypsithermal, a postglacial time of significantly warmer temperatures. During this period, the grasslands extended further eastward in the so-called "prairie peninsula" (page 166). A few small prairies are found on well-drained, warm sites in the western half of Ohio. Oak-Hickory stands are also thought to have been more widespread during the Hypsithermal (page 64); today, they occur in the Beech-Maple region on coarse, dry soils.

Maple-Basswood Association.   This covers the smallest area of the deciduous forest associations. Its region includes parts of eastcentral and southeast Minnesota, southwestern Wisconsin, and northeast Iowa. All of its boundaries are indistinct; some authorities have claimed that this is partly a result of Indian burning practices.

The dominant species are sugar maple and American basswood *(Tilia americana)*; oaks predominate among the subdominants. Despite its northerly location, the Maple-Basswood association is not entirely postglacial in origin. Southwestern Wisconsin is unglaciated and makes up most of the "Driftless Area" (a name that refers to the absence of glacial deposits). Glaciers passed north, east, south, and west of this area, but it was never completely surrounded by ice at any one glacial stage. The Driftless Area is thought to have been a refugium for a variety of species, possibly including some of those of the Maple-Basswood association.

Hemlock-White Pine-Northern Hardwoods Association.   Vegetation scientists have a range of views on the significance of this association. Some consider it just a broad transition between the true deciduous and Boreal coniferous forests. Others believe it to be so unique that they classify it separately from the deciduous and Boreal forests. Still others treat it as an association of the deciduous forest, as is done here. Its range is from western Ontario and Minnesota across southern Canada and northern United States to the Atlantic Coast. Similar vegetation may be found below the coniferous forest zone of the Appalachian Mountains.

FIGURE 9.7 Vegetation diversity in the Hemlock-White Pine-Northern Hardwoods association along the Ontario shore of Lake Superior. The deciduous species appear lighter than the conifers.

Much of the association consists of a mosaic of stands—some dominated by deciduous species, some by coniferous species, and others with a proportion of both (Figure 9.7). This diversity is why some scientists consider it a transitional type of vegetation. The soils also are varied, with both Podzol and Podzolic soils present. Species with distributions largely centered in this forest region include white pine *(Pinus strobus)*, red pine *(Pinus resinosa)*, and yellow birch *(Betula lutea)*. The dominance of these species provides the basis for other scientists to classify this association separate from the deciduous and coniferous forests. Other dominants have Boreal forest affinities. These include paper birch *(Betula papyrifera)*, quaking aspen *(Populus tremuloides)*, and jack pine *(Pinus banksiana)*. Most of the dominants, however, have a deciduous forest distribution. Examples include red maple *(Acer rubrum)*, sugar maple, American beech, American basswood, and eastern hemlock.

Stands of deciduous species are dominated by sugar maple, American beech, American basswood, birches, and quaking aspen. Stands of coniferous species are characterized by white, red, and jack pines on dry sites and by such Boreal forest species as white spruce *(Picea glauca)*, black spruce *(Picea mariana)*, and tamarack or larch *(Larix laricina)* in scattered bogs. Red spruce *(Picea rubens)* is a widespread dominant of eastern coniferous stands. Mixed stands frequently contain the deciduous species and white pine, eastern hemlock, or red spruce; white pine occurs throughout the range of the association, hemlock is primarily in the east, and red spruce is only in the east.

The mosaic pattern of this association is related to several factors. Varied soil is one factor; in general, the sandy, nutrient poor soils support stands of conifers, whereas the more fertile soils have hardwoods. Fires also may have an effect; in some areas the deciduous species are considered to be seral to the conifers, but in others the reverse is thought to be true.

Prior to European settlement, much of this region was dominated by huge white pines, many around 65 meters in height. The pines are seral species that were formerly maintained by light surface fires. The cutting of these forests for timber began as early as 1623 in the east, peaked in the middle of the nineteenth century in the Great Lakes region, and rapidly declined shortly thereafter. Extensive slash fires followed logging and destroyed white pine regeneration. Without sufficient seed sources for the pines, much of the region was revegetated by hardwoods, nearly all of which root sprout. Today, vast areas of hardwoods exist where there had been conifers. These sprout forests are a reflection of past land abuse.

Oak-Chestnut Association.   This association no longer exists in its original composition. Its range included all but the northern hardwoods and coniferous forest zones of the Appalachian Mountains and adjacent, paralleling physiographic regions. In the north, the Oak-Chestnut association extended through the lower elevations of most of Pennsylvania and southern New England. The dominant climax species of the presettlement forest were American chestnut *(Castanea dentata)*, chestnut oak *(Quercus prinus)*, red oak, and several other oaks. Pitch pine *(Pinus rigida)* was an important seral species throughout the range. Heath shrubs, especially azaleas and rhododendrons (both *Rhododendron* spp.), were characteristic of the understory. All of these species, with the exception of chestnut, continue to be important.

Chestnut blight was introduced into the United States from China around the turn of the century. The fungus which causes the blight was first reported in 1906 in New York City, and in seven years it had covered all of New Jersey and parts of seven other states. It destroyed chestnut trees by killing the phloem (food conducting) layer in the trunks. Attempts to control the spread of the blight proved futile. By 1920, the blight had killed 50 percent of the chestnut trees as far south as Maryland and as far west as portions of Pennsylvania; by 1930, 50 percent had died south into North Carolina and west into Ohio; today, few trees are left anywhere (Figure 9.8). In many cases, the root systems were not killed, so sprouts have continued to form, but these have been repeatedly destroyed by the blight.

The loss of the American chestnut was important not only biologically, but also economically. It was the most versatile tree in North America. Hot,

FIGURE 9.8   The trunk of a dead chestnut killed by chestnut blight in Great Smoky Mountains National Park, Tennessee.

roasted chestnuts were sold for food throughout the east; its wood was used for construction, paneling, trim, furniture, telephone poles, fences, shingles, and for many other items; and industry relied on it for chemicals used in tanning leather. It accounted for one-fourth of all the hardwood timber cut for lumber in the southern Appalachians.

Today, the forest could be called the Oak association. Infected chestnut trees died slowly, so there was a gradual increase in the importance of the other dominants, instead of invasion of new species. Nevertheless, it will take some time before stands reach equilibrium.

Oak-Hickory Association.   This is the most widespread of the deciduous forest associations. Its distribution is U-shaped, extending from New Jersey south along the Piedmont physiographic region into the Gulf Coast states, west to central Texas, and north to Minnesota. In the east and south, mature Oak-Hickory stands are not common, but where they occur their structure is well developed, resembling the stands of other central deciduous forest associations. In the west where precipitation is lower, the size and density of the trees decreases, and the forests are more open (Figure 9.9). At its western edge, the Oak-Hickory association grades into the grassland as a savanna in the north and south (Chapter 12) and in a mosaic pattern from eastern Kansas to western Indiana, the location of the prairie peninsula.

Throughout the range of the association, mature stands are dominated by

FIGURE 9.9   A stand of Oak-Hickory vegetation in eastern Oklahoma. The dominant is post oak.

oaks, with hickories secondary in importance. Some of the more widespread oaks are white oak, red oak, and black oak *(Quercus velutina)*. In addition, bur oak *(Quercus macrocarpa)* is common in the west, especially the northwest, and post oak *(Quercus stellata)* and blackjack oak *(Quercus marilandica)* are important in the south and southwest. Common hickories include bitternut hickory *(Carya cordiformis)*, mockernut hickory *(Carya tomentosa)*, red hickory *(Carya ovalis)*, and shagbark hickory *(Carya ovata)*. Greatest species diversity occurs in the low mountains of central Arkansas.

As mentioned before, Oak-Hickory stands are not common in the east and south. Instead, pine or pine-hardwood stands predominate (Figure 9.10). Therefore, the vegetation of this region can be classified as Oak-Pine or Oak-Hickory-Pine to emphasize the importance of conifers. The Piedmont region centered around Georgia serves as an example of this region. Its presettlement vegetation consisted of 35 to 40 percent hardwood stands (primarily Oak-Hickory), 45 percent mixed pine-hardwood stands, and 15 percent pine stands. The pines, considered to be seral species, were maintained by lightning strike and Indian fires.

Settlement of Georgia's Piedmont began in 1773 and, by 1840, 87 percent of the region was under cultivation, with cotton as the dominant crop. Intensive land use resulted in great soil erosion and decreased fertility. An estimated 10 percent of the farmland was abandoned with the loss of labor during the Civil War, 30 percent with the agricultural depression of the late

1880s, and 35 percent following the spread of the boll weevil in the 1920s. Succession on abandoned land involved rapid revegetation by pines.

Throughout the southern and eastern Oak-Hickory region, pines are especially suited to the infertile, porous soil. In the south, loblolly pine *(Pinus taeda)* and shortleaf pine *(Pinus echinata)* predominate. In the north, pitch pine and Virginia pine *(Pinus virginiana)* are most common.

With reforestation of much of the abandoned land in the South, timber production became a major land use. Originally, the wood was cut into lumber, but today most of it is used for pulpwood. Naval stores (e.g., turpentine) are also an important product. Pine forests are now an agricultural crop of the South; they are grown systematically, using genetically improved stock. The ten species of southern pines (including those more common in the Southern Mixed Hardwoods region) presently account for 25 percent of the United States timber supply east of the Rocky Mountains.

The key environmental factor in maintaining these pine stands is fire. Without it, succession leads to replacement by hardwood species, primarily oaks and hickories (Figure 9.10). Fires in mixed stands kill most hardwoods but not the thicker barked pines. Fires also produce a mineral seedbed that is ideal for the germination of pine seeds. Other fire adaptations include cone serotiny in pitch pine in areas of high fire frequency (Figure 9.11). Today, prescribed burning is used to maintain the pine stands, as well as to reduce fuel loads and thereby lower the danger of destructive wildfires.

FIGURE 9.10 A pine-dominated stand of Oak-Hickory vegetation in southwestern Tennessee. The dominant pine is loblolly pine. Note the development of deciduous species in the understory and to the left.

FIGURE 9.11   A serotinous cone of pitch pine in southern New Jersey that was opened by fire a few hours before. Near the center of the photograph are light-colored, winged seeds that have fallen out of the cone.

Southern Mixed Hardwoods Association.   This association occurs in the Coastal Plain physiographic region from North Carolina to Texas, excluding the tip of Florida. Since some of its dominants are evergreen angiosperms, this association is sometimes considered to be transitional between Mixed Mesophytic and tropical forests. However, its overall composition is much more like the Mixed Mesophytic forest, so it is included with the deciduous forest formation.

The Coastal Plain region has many diverse types of vegetation, including freshwater swamps, pine woods, and coastal vegetation, but the association is named after the climax community of upland areas (Figure 9.12). Even within this community, individual stands may show great differences. Dominant species include American beech, white oak, live oak *(Quercus virginiana)*, laurel oak *(Quercus laurifolia)*, evergreen magnolia *(Magnolia grandiflora)*, and approximately 10 others. Mature stands may have five to nine codominants. Wet lowlands have several kinds of freshwater swamp communities. Sites that are continuously wet or nearly so are dominated by bald cypress *(Taxodium distichum)*, a deciduous gymnosperm (Figure 9.13). Pocosins or bogs of evergreen shrubs and small trees are another type of vegetation found in permanently wet sites. Areas of seasonally fluctuating water levels have either deciduous or evergreen hardwoods; the former are on the more fertile sites. A common plant found on the limbs of trees in many of these communities is Spanish moss *(Tillandsia usneoides)*, an angiosperm epiphyte (Figure 9.13). It has scales that absorb moisture, serving the same purpose as the roots of terrestrial plants.

The pine woods are fire-dependent seral communities. The most widespread dominant is longleaf pine *(Pinus palustris)*, but loblolly pine and slash pine *(Pinus elliottii)* are also important. Many of the communities are open stands with a dense herbaceous cover. Before settlement, the pine woods were maintained by fire (and Indian agriculture which periodically left fields abandoned). Following settlement, there were alternating periods of lumbering, farming, and land abandonment. Eventually pines came to be treated as an agricultural crop, and their relationship with fire was studied.

Longleaf pine, for example, is favored by fire since burning exposes mineral soil, reduces competition, and partially controls brown-sprout needle disease—a fungus that can defoliate young pines. Longleaf pines are adapted to fire because of their growth pattern and growth-form. Young seedlings show little vertical growth for approximately three to six years or more; only the terminal bud and a tuft of needles is formed (Figure 9.14). Net production goes into the development of a taproot that, at this stage, will grow to a meter or more in length. During this "grass stage," the plants are unlikely to be killed by fires. This stage of growth ends when the pine seedlings show rapid vertical growth. Fire may kill the trees before sufficiently thick, protective bark has been formed; however, natural stands contain individuals in various stages of growth, so periodic fires do not eliminate all the pines. In contrast, the aerial portions of all young hard-

FIGURE 9.12 A stand of Southern Mixed Hardwoods vegetation near Charleston, South Carolina. The dominant is live oak.

FIGURE 9.13   Bald cypress near Charleston, South Carolina. Note its "knees"—woody structures protruding from the water. Spanish moss, an epiphyte, is hanging from the tree's branches.

woods are killed; therefore, in stands managed for wood production, periodic prescribed burning is used to remove hardwood competitors.

The fire dependency of species in this region and elsewhere has led to the hypothesis that such species are actually adapted to increasing the probability of fire. For example, longleaf pine has lengthy needles that fall, catch on shorter plants, and increase the fuel load around these competitors (Figure 9.15).

Flood-Plain Forests.   These forests are not classified as a separate association, but they have an extensive distribution along rivers throughout the deciduous forest. Such riparian forests are subject to periodic flooding, and sediments provide a major source of mineral nutrients. Soil moisture is generally high, and species characteristic of these forests are adapted to poor soil aeration.

FIGURE 9.14 The "grass stage" of growth in the longleaf pine in eastern South Carolina.

FIGURE 9.15 Saplings and shrubs of deciduous species (leafless in early spring) that are covered with needles from the canopy pines. This increases the probability of these plants being killed in a fire. Location: southern South Carolina.

The flood-plain forests of eastern North America can be divided into two parts. Southern forests are dominated by oaks, bald cypress, and tupelos (*Nyssa* spp.), as are the freshwater swamps of the Southern Mixed Hardwoods association. Northern flood-plain forests are dominated by cottonwoods (*Populus* spp.), willows (*Salix* spp.), elms (*Ulmus* spp.), and sometimes ashes. These forests extend into the grassland formation, since high soil-moisture levels compensate for low amounts of precipitation. Other types of deciduous riparian forests occur further west, for example, in California.

Adaptations

The dominant characteristic of the climate in the deciduous forest region is the division of the year into distinct seasons, including a winter with continuous below-freezing temperatures in the north and at least occasional frosts in the south. Since the broad leaves of angiosperms are ill-suited for withstanding winter conditions, the deciduous habit is highly adaptive.

The timing of autumn leaf fall is largely determined by day lengths, although other climatic factors such as late summer or early fall drought also may be involved. As day lengths shorten, the critical photoperiod is sensed by a pigment in the plant, and a hormone is produced that stimulates the weakening of cell walls at the base of the leaves and initiates the formation of bark where the leaves are attached. Thus, since the buds have strong thick scales, the entire tree or shrub is protected against rapid transpiration in the winter when low soil temperatures retard or prevent root absorption. Nevertheless, some transpiration does occur. Deciduous angiosperms generally have higher rates of winter water loss than evergreen conifers—a factor that is most likely important in their respective distributions.

Low winter temperatures also pose a problem. Resistance to frost damage is achieved by hardening. This process is not fully understood, but it begins in the early fall and involves the development of cellular dormancy, a relatively inactive state. Frost (and drought) resistance is thought to be related to a great increase in the sugar content of the cell sap.

Spring growth patterns are very important in the deciduous forest. Herbs are the first to show signs of growth. As in the tundra, virtually all herbs within an undisturbed deciduous forest are perennials, a growth-form that is adaptive to short growing seasons. Many of these herbs grow, flower, and produce seed in the relatively short period before the leaves of the canopy trees fully develop.

Perennial herbs, since they sprout from underground storage organs such as bulbs, can reach full size faster than annuals. Rapid spring growth in perennials is aided by the shoots being preformed the previous fall, though this probably makes them more susceptible to winter frost injury. The length of time for emergence in the spring is inversely proportional to the length of the winter cold period (i.e., the longer the cold period, the shorter the time for emergence). This ensures against growth during a few brief warm days in the late fall or winter; it also provides for rapid growth in the spring. The high sugar levels of winter may be maintained throughout the spring; however, photosynthesis leads to rapid accumulation of starch, indicating that most of the net production is stored. Many spring herbaceous perennials do not carry on photosynthesis throughout the summer, but become dormant as aerial parts die following closure of the canopy.

Initiation of spring growth in woody plants also is thought to be primarily under the control of temperature. It has been suggested that some trees form leaves later than herbs because the deeper roots of trees are warmed later than the shallow underground organs of the herbs.

The leaf form of most deciduous forest plants has great surface exposure—an obvious benefit in photosynthesis as long as soil moisture is adequate to replace transpirational losses. Leaf morphology varies greatly. Even on an individual tree, "sun" and "shade" leaves can be described. Sun leaves are those exposed to high light intensity on the outside of the crown of a tree. They are generally smaller, thicker, hairier, and more deeply lobed than the shade leaves of the interior of the crown. Leaves that are broad, such as shade leaves, when exposed to bright light in still air may be 10 to 20° C warmer than the air temperature and, therefore, may approach the killing thermal unit of 55 to 65° C. The smaller surface area and deeper lobes of sun leaves are adaptive, since they reduce light absorption and increase heat dissipation by wind. Also, their greater thickness results in better conduction of heat from the center to the outer edge. In white oak, S. Vogel has reported that the difference between air and leaf temperatures is 20 percent less for sun leaves than shade leaves exposed to the same amount of light in still air.

The arrangement of sun and shade leaves is also adaptive. Shade leaves are typically oriented at right angles to the light. Sun leaves are usually inclined; thus, their heat load and water loss are reduced and light penetration into the crown of the tree is increased. Some physiological differences between sun and shade leaves include the ability of shade leaves to pass the compensation point (where respiration equals photosynthesis) and reach maximum photosynthesis at lower light intensities than sun leaves.

Many of the differences between sun and shade leaves also appear in canopy versus understory plants and in plants of dry habitats versus those of mesic habitats.

A large number of adaptations of deciduous forest plants relate to succession. One general observation made by H. S. Horn is that early successional species have multilayered leaf arrangements and climax species have monolayered arrangements. The multilayered growth-form has several layers of leaves; it is adaptive under conditions of high light intensities, since even the inner leaves of a tree are above the compensation point. However, the low light intensity beneath a dense canopy of a late seral stage would result in these leaves being below the compensation point and, consequently, being energy drains on the tree. Hence, the monolayered growth-form (with a single layer of leaves) is adaptive in later stages of succession.

In a dense forest, it would be advantageous for a tree to have a high degree of phenotypic plasticity to enable it to survive under suppressed conditions beneath the canopy and later grow into it (either by gradual increases in height or rapid development following the formation of a gap in the canopy). The ability to control leaf structure and, theoretically, leaf arrangements are possible means for a species to be successful both in the understory and in the canopy.

Crown shape and limb characteristics are also thought to be important in relation to ice storms. It has been suggested that in the northeast United States, where ice storms are frequent, they cause greater damage to seral species than climax species and therefore speed the rate of succession. In contrast, some seral deciduous forest species may be adapted to slowing the rate of succession by producing chemical exudates that inhibit the growth of understory plants.

Human Impact

The effects of the aboriginal Indians on the deciduous forest appear to have varied greatly from region to region. For example, their impact was small in the mid-Appalachians but was large in the northeast and west. Early European explorers reported many fires caused by Indians along the Atlantic Coast. As a result of a high fire frequency, the presettlement forests of the northeast were open and parklike. Locally, the Indians cleared land for villages and agricultural fields; the range of this effect was broadened by seasonal migrations and village relocations. Indian tribes near the western boundary of the deciduous forest set fires that kept the western Oak-Hickory forest very open, helped determine the grassland-forest boundary (page 220), and reportedly limited the range of the Maple-Basswood association by maintaining xeric conditions more suited to Oak-Hickory vegetation.

European settlement resulted in a rapid clearing of forest land in the northeast for agriculture. The area of agricultural land was at a maximum between 1820 and 1850; many counties were 75 to 80 percent cleared. Soil deterioration and the opening of areas of more fertile land further west led to field abandonment and reforestation. Nevertheless, cutting to clear land remained a widespread practice throughout the region; environmental conditions that favor the development of deciduous forest vegetation are also good for agriculture and grazing.

New technological devices, such as portable steam-powered sawmills, made possible the development of a major logging industry. The forest was cut for lumber, pulpwood, and fuel. Fuel was especially important prior to the development of coal mining in Kentucky and West Virginia. Reforestation took place, especially as the grassland region developed into an agricultural belt; however, in large areas the logging was followed by slash fires that prevented the regeneration of forests that had been typical of the region. An example of this is the replacement of coniferous stands of the Hemlock-White Pine-Northern Hardwoods forest by a hardwood sprout forest.

Today, remnants of the original deciduous forest associations are widely scattered in a few parks and on private property. Nearly all of the land that is still forested is second, third, or even fourth growth (Figure 9.16). The total area of the forest is being reduced continually by phenomena such as suburban sprawl and strip mining.

FIGURE 9.16   A dense second growth forest in Great Smoky Mountains National Park, Tennessee. The stones in the center of the photograph are the remains of a wall around a farmer's field.

SUGGESTED READINGS FOR FURTHER STUDY

Boardman, N. K. 1977. Comparative photosynthesis of sun and shade plants. Annual Review of Plant Physiology 28:355–377.

Braun, E. L. 1950. Deciduous forests of eastern North America. Blakiston, Philadelphia. 596 p.

Bray, J. R. 1956. Gap phase replacement in a maple-basswood forest. Ecology 37:598–600.

Cain, S. A. 1935. Studies on virgin hardwood forest: III. Warren's Woods, a beech-maple climax forest in Berrien County, Michigan. Ecology 16:500–513.

———. 1943. The Tertiary character of the cove hardwood forests of the Great Smoky Mountains National Park. Bulletin of the Torrey Botanical Club 70:213–235.

Curtis, J. T. 1959. The vegetation of Wisconsin: An ordination of plant communities. University of Wisconsin Press, Madison. 657 p.

Hepting, G. H. 1974. Death of the American chestnut. Journal of Forest History 18:60–67.

Horn, H. S. 1971. The adaptive geometry of trees. Princeton University Press, Princeton, New Jersey. 144 p.

Johnson, F. L. and P. G. Risser. 1972. Some vegetation-environment relationships in the upland forests of Oklahoma. Journal of Ecology 60:655–663.

Komarek, E. V. 1974. Effects of fire on temperate forests and related ecosystems: Southeastern United States, p. 251–277. *In* T. T. Kozlowski and C. E. Ahlgren (eds.). Fire and ecosystems. Academic Press, New York.

Little, S. 1974. Effects of fire on temperate forests: Northeastern United States, p. 225–250. *In* T. T. Kozlowski and C. E. Ahlgren (eds.). Fire and ecosystems. Academic Press, New York.

Mackey, H. E., Jr. and N. Sivec. 1973. The present composition of a former oak-chestnut forest in the Allegheny Mountains of western Pennsylvania. Ecology 54:915–919.

Martin, W. H. 1975. The Lilley Cornett Woods: A stable mixed mesophytic forest in Kentucky. Botanical Gazette 136:171–183.

McIntosh, R. P. 1972. Forests of the Catskill Mountains, New York. Ecological Monographs 42:143–161.

Monk, C. D. 1965. Southern mixed hardwood forest of northcentral Florida. Ecological Monographs 35:335–354.

Oosting, H. J. 1942. An ecological analysis of the plant communities of Piedmont, North Carolina. American Midland Naturalist 28:1–126.

Penfound, W. T. 1952. Southern swamps and marshes. Botanical Review 18:413–446.

Quarterman, E. and C. Keever. 1962. Southern mixed hardwood forest: Climax in the southeastern Coastal Plain, U.S.A. Ecological Monographs 32:167–185.

Rice, E. L. and W. T. Penfound. 1959. The upland forests of Oklahoma. Ecology 40:593–608.

Robichaud, B. and M. F. Buell. 1973. Vegetation of New Jersey: A study of landscape diversity. Rutgers University Press, New Brunswick, New Jersey. 340 p.

Vankat, J. L., W. H. Blackwell, Jr., and W. E. Hopkins. 1975. The dynamics of Hueston Woods and a review of the question of the successional status of the southern beech-maple forest. Castanea 40:290–308.

Vogel, S. 1968. "Sun leaves" and "shade leaves": Differences in convective heat dissipation. Ecology 49:1203–1204.

Whitehead, D. R. 1973. Late-Wisconsin vegetational changes in unglaciated eastern North America. Quaternary Research 3:621–631.

Whittaker, R. H. 1956. Vegetation of the Great Smoky Mountains. Ecological Monographs 26:1–80.

Whittaker, R. H. and G. M. Woodwell. 1969. Structure, production and diversity of the oak-pine forest at Brookhaven, New York. Journal of Ecology 57:155–174.

10

GRASSLAND VEGETATION

The grassland formation extends over the vast central part of North America and smaller regions further west. Worldwide, grasslands cover nearly one-fourth of the land surface; the Russian steppe, the South African veld, and the South American pampas are familiar examples. Yet the French explorers who were the first Europeans to see the central North American grasslands had never observed anything like it before. Lacking an adequate word for this vegetation, they used the term "prairie," which meant meadow. Even today, the easternmost portion of the grasslands, that first seen by the French, is considered to be the "true prairie."

In comparison to forests, most grasslands appear to have a rather simple structure; however, study shows greater complexity than might be expected. Two herb layers are always recognizable, and some grasslands have three. In addition, there is a surface layer of mosses and lichens. Grasslands also have stratification of root systems, with different species dominating particular soil depths. Again, there may be two or three layers.

DISTRIBUTION

The distribution of the grassland formation clearly relates to its origin, since grasslands evolved with the development of rain shadows following mountain uplifting in western North America (page 61). The central grasslands extend from Indiana westward to the Rocky Mountains and from southcentral Canada to portions of northern Mexico (Figure 10.1). There is great variation in structure and composition over this broad range.

A simple classification divides the central grassland into the Tall Grass,

Mixed Grass, and Short Grass associations from east to west. The boundaries between these associations are relatively indistinct and vary with climatic changes. Other associations are the Desert Grassland in portions of Arizona, New Mexico, Texas, and northern Mexico, the Palouse Prairie in parts of Washington, Oregon, and Idaho, and the California Grassland. With this wide distribution, the grasslands come into contact with the coniferous forest, deciduous forest, desert, shrubland, and woodland formations.

ENVIRONMENT

The climate of grasslands is characterized by a low precipitation–evaporation ratio (Figure 10.2). Only the climates of deserts have lower ratios. Mean annual precipitation values in the central grasslands decrease from east to west and from south to north. Typical values are 100 centimeters in the southeast and 65 centimeters in the northeast. Further west in the Mixed Grass association, values are 65 centimeters in the south and 35 centimeters in the north. In the Short Grass association at the base of the Rocky Mountains, precipitation values average around 25 centimeters per year. The Desert Grassland is especially dry.

A high proportion of the rainfall in the central grasslands occurs during the growing season; maximum values are in June in the east and somewhat earlier in the west. Late summer and fall precipitation is usually small. Droughts result as this low rainfall combines with high temperatures and drying winds. In addition, droughts lasting throughout the growing season and often extending over several years may develop periodically.

The location of the central grasslands results in a continental climate; temperatures are hot in the summers and cold in the winters. Extremes of $45°$ C and $-35°$ C are not uncommon in the center of the range. Evaporation rates are high throughout the year; on an annual basis potential evaporation far exceeds precipitation. The soils rarely remain moist throughout the summer, so the lengths of the growing seasons are limited by cold winters and late summer drought. Growing seasons average from 120 to 200 days from north to south in the Mixed Grass region.

The climates of the Palouse Prairie and California Grassland regions are for the most part very similar to that of the central grasslands (Figure 10.2). The major difference is in the timing of precipitation. With the Mediterranean-type climate of the far west, these two grasslands have wet winters but little precipitation during the summers. Average annual precipitation ranges from 15 to 50 centimeters.

The soils of the grassland formation reflect variations in climate and vegetation. From east to west across the central grasslands there are the

(a) Tall Grass vegetation in southern Wisconsin.
(b) Flowering head of a grass.

(*c*) Sunflowers in Tall Grass vegetation in northeast Illinois.
(*d*) Mixed Grass vegetation in central Kansas.

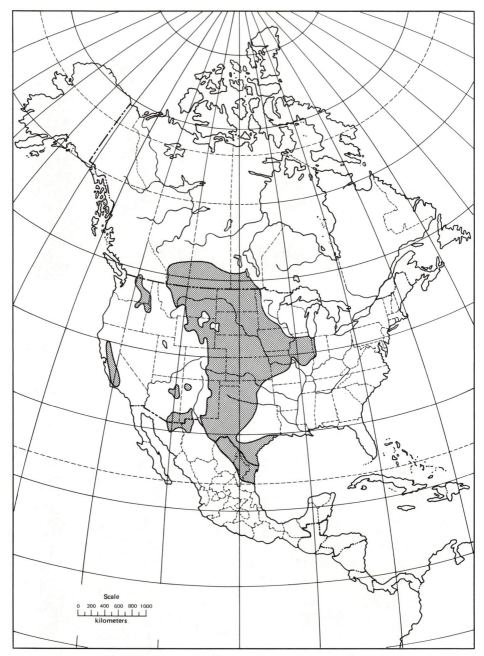

FIGURE 10.1 Distribution map of the grasslands.

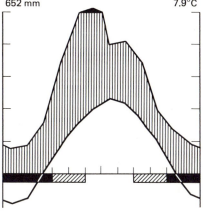

Sioux Falls, South Dakota (435 m)

652 mm   7.9°C

North Platte, Nebraska (856 m)

468 mm   9.1°C

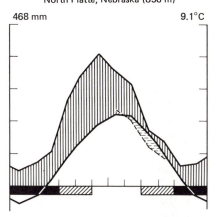

Pueblo, Colorado (1461 m)

312 mm   10.9°C

Albuquerque (airport), New Mexico (1620 m)

205 mm   13.0°C

Dayton, Washington (519 m)

507 mm   10.6°C

Fresno, California (101 m)

231 mm   17.2°C

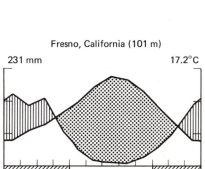

FIGURE 10.2 Climate diagrams for the grassland region: Sioux Falls, North Platte, and Pueblo are in the Tall, Mixed, and Short Grass associations respectively; Albuquerque is in the Desert Grassland; Dayton is in the Palouse Prairie; Fresno is in the California Grassland. *Source.* Redrawn from Walter, H. and H. Lieth. 1967. Klimadiagramm-weltatlas. Fischer, Jena, East Germany.

following Great Soils: Prairie, Chernozem, and Brown and Chestnut. These correspond approximately to the regions of the Tall, Mixed, and Short Grass associations. The Prairie Great Soil type is among the most fertile in the world. It has a high amount of organic matter that is readily broken down by decomposers. The moderate amounts of precipitation result in less leaching of nutrients than occurs in the Podzolic soils of the deciduous forest.

The Prairie soil is a pedalfer, and the Chernozem and Brown and Chestnut types are pedocals. These Mixed and Short Grass soils are not as well leached as the Prairie soil; however, since they have less organic matter, they are lower in fertility. The profile of these pedocal soils includes a layer of calcium carbonate that is formed as precipitation infiltrates the soil, dissolves calcium, leaches it, and deposits it at the depth of leaching. The position of the layer therefore reflects the amount of precipitation. It is usually around 35 centimeters deep in Brown and Chestnut soils and 35 to 150 centimeters in Chernozem soils.

Low precipitation–evaporation ratios result in the increased importance of topographic relief. The elevation of the central grasslands generally increases from the Mississippi River drainage to the foothills of the Rocky Mountains. Throughout the region, the relief is flat to rolling. Low areas are frequently more mesic than slopes or hilltops. The composition of grassland communities changes along such soil-moisture gradients.

Fire is a very important, if not essential, ecological factor in grasslands (page 175; Figure 4.10). Lightning fires were always part of the environment. The frequency of fires increased following the human occupation of North America, since Indians intentionally and accidentally set fires. The majority of fires occurred in the fall, when standing biomass was greatest and the environment was driest. Strong winds and the lack of natural barriers meant that fires commonly burned large areas at great speeds. The front of one fire in the early part of this century covered 200 kilometers in a single day.

Even large fires were less intense than fires in forests, since there is less fuel in grasslands. Soil-surface temperatures rarely reach much over 100°C in grassland fires, so the direct effects on soil structure and chemistry are minimal. Fires do not kill most grassland plants, but they do destroy woody plants (page 46). There is little fertilization from ash following grassland fires; however, the darkened soil surface causes earlier and greater warming of the soil in the spring. This favors the growth of early season plants and may increase the activity of nitrogen-fixing bacteria. Plants in burned areas typically produce shoots with higher protein and mineral levels than plants in unburned areas. Fire also stimulates the overall growth of some grassland species in ways that are not yet known.

The biotic factor is also an important aspect of the environment of grasslands. An example was given in Chapter 4 (page 45).

FIGURE 10.3   Grass sprouts form from a rhizome (an underground stem), exposed here by erosion.

VEGETATION

The dominant growth-form is obviously the grass type; its aerial growth takes place from a point at or beneath the soil surface (Figure 10.3). Forbs are also very widespread, especially members of the sunflower (Compositae; photograph *c*, page 161) and legume families (Leguminosae). Most of the grasses and forbs are perennials. Annuals account for fewer than 5 percent of the species in the relatively undisturbed areas of the central grasslands.

The perennial grasses can be divided into two major types: sod grasses and bunch grasses. Sod grasses reproduce both by seeds and by vegetative means. They frequently have rhizomes (underground stems) or, in a few species, stolons (prostrate above-ground stems, i.e., "runners"). These structures produce a series of new aerial shoots as they spread away from the parent plant. Consequently, sod grasses form mats that can be similar in density to suburban lawns. Bunch grasses, on the other hand, only have erect stems. Single plants may form a hundred or more shoots, but these do not spread as mats; instead, they form dense clumps of aerial stems. Some grass species may vary in their growth-form, developing as sod grasses in certain environments and as bunch grasses in others.

Grassland vegetation is very diverse. The classification into the six associations mentioned earlier in this chapter is fairly well agreed upon; however, there is no consensus on the subdivisions of these associations. Grassland vegetation also shows great seasonal variation; 17 new species come into flower each week from April through October in one Wisconsin prairie (photograph *a*, page 160). Therefore, studies of species dominance done early in the growing season differ from those done later. The following section does not concentrate on either regional or seasonal variations, but presents an overall picture of the six major associations.

Associations

Tall Grass.   Late in the growing season, when the plants of this association have reached their full height, many of them have stems of two or more meters (Figure 10.4). The early settlers of the region wrote of men on horseback who seemingly bobbed along the surface of the prairie grasses, their mounts hidden by the tall plants.

Before settlement, this association bounded the deciduous forest; its range was narrow in the north and south, but wide where it extended eastward in the prairie peninsula of Illinois. The existence of this lobe of Tall Grass prairie into the deciduous forest has been attributed to the climatic pattern of North America. Westerly winds coming across the Rocky Mountains are modified by the Arctic air mass to the north and by the tropical Atlantic air mass to the south. The dry Westerlies extend furthest eastward

FIGURE 10.4   Tall Grass vegetation in southern Wisconsin dominated by big bluestem and goldenrod (*Solidago* spp.) in late August.

between these two air masses and produce a climate that can support either grassland or forest in the prairie peninsula region. Reoccurring fires maintained the grassland vegetation in this area of borderline climate, just as fires further north and south were also the main factor in determining the position of the grassland-forest boundary (page 220).

Early in the growing season, with the smaller species predominating and the others just beginning growth, the tall aspect of this association is not apparent. Throughout the growing season there is a progression of plant species to greater heights. In fall, stratification is more evident than in any other grassland association; typically, there are three herb layers with the majority of dominant species in the tallest stratum.

This association also has the greatest species diversity. Dominants on lowland sites include big bluestem (*Andropogon gerardi;* photograph *b,* page 160), Indian grass *(Sorghastrum nutans),* switchgrass *(Panicum virgatum),* and Canada wild-rye *(Elymus canadensis),* with prairie cordgrass *(Spartina pectinata)* common in wet areas of poor soil aeration. The dominants of upland, drier areas include little bluestem *(Andropogon scoparius),* needlegrass *(Stipa spartea),* prairie dropseed *(Sporobolus heterolepis),* side-oats grama *(Bouteloua curtipendula),* and Junegrass *(Koeleria cristata).* The above species include a mixture of bunch and sod grasses and grasses of various heights. Most of the grasses are of southern origin and consequently do not flower until late summer.

Mixed Grass.   This association lies between the Tall and Short Grass regions and derives its name from the fact that it contains a mixture of both groups of species. The two most noticeable strata are a dense layer of grasses at a height of about 30 centimeters and a more open layer of midgrasses at approximately 125 centimeters (Figure 10.5 and photograph *d,* page 161). The shorter dominants include blue grama *(Bouteloua gracilis),* hairy grama *(Bouteloua hirsuta),* and buffalo grass *(Buchloë dactyloides).* Taller grasses include little bluestem and needle-and-thread grass *(Stipa comata).*

The eastern and western boundaries of the Mixed Grass association continually vary according to amounts of precipitation. A series of dry years results in an increased dominance of short grasses, since they are generally better adapted to a dry climate. Thus, under these conditions the boundaries of the Mixed Grass region shift eastward. Westward shifts occur after several years of relatively high precipitation, which favors the taller grasses.

Short Grass.   This westernmost portion of the central grasslands occupies the "High Plains" and is composed chiefly of grasses 20 to 50 centimeters in height (Figure 10.6). Many of them are sod grasses, whose shallow root systems are well adapted to the dry conditions. Peak growth occurs early in

FIGURE 10.5 Mixed Grass vegetation in southwest Kansas dominated by little bluestem.

FIGURE 10.6 Short Grass vegetation in eastern Colorado dominated by needle-and-thread grass.

the summer. By the time soil moisture is gone in mid to late summer, many of the plants have already ceased photosynthesis and become dormant.

The major dominants are buffalo grass and blue grama, both of which extend eastward well beyond the Short Grass region. Other important species include needle-and-thread grass, wire grass *(Aristida longiseta)*, and western wheat grass *(Agropyron smithii)*; these form an open layer above the dominant short grasses, but grazing reduces their height. Prior to the middle of the nineteenth century, vast herds of buffalo and smaller numbers of elk and antelope were the primary grazers. Today, cattle produce the short grass appearance.

Desert Grassland.   This association is scattered across the southwest United States and northern Mexico. It is most frequently found at elevations between the Warm Desert and the Pinyon-Juniper woodland (Chapters 11 and 12). Dominant grasses include grama grasses *(Bouteloua* spp.), tobasa grasses *(Hilaria* spp.), and wire grasses *(Aristida* spp.). Shrubs and other growth forms were always present but have become very widespread since pioneer settlement (Figure 10.7; page 177). Mesquite *(Prosopis juliflora)*, a shrub, is the most common of these, but others include creosote bush *(Larrea divaricata)*, acacias *(Acacia* spp.), yuccas *(Yucca* spp.), and prickly pear cacti *(Opuntia* spp.). Except for the common occurrence of these plants, which are primarily Hot Desert species, the Desert Grassland is closely related to the southern Short and Mixed Grass associations.

FIGURE 10.7   Desert Grassland vegetation in southwest New Mexico with yucca, a tall semisucculent species (page 192).

FIGURE 10.8   Palouse Prairie vegetation in Idaho.

Palouse Prairie.   This association is found in the rain shadow of the Cascade Mountains, that is, in western Washington and Oregon and eastern Idaho (Figure 10.8). With a Mediterranean-type climate, both the Palouse Prairie and the California Grassland have some growth in the fall, a small amount of growth during the winter (when low temperatures are limiting), and a burst of growth in the spring. By midsummer the plants have browned and become dormant, while the eastern central grasslands are still green.

The dominant grasses of the Palouse region used to be bluebunch wheat grass *(Agropyron spicatum)*, Idaho fescue *(Festuca idahoensis)*, and sandberg bluegrass *(Poa secunda)*. Today, downy chess *(Bromus tectorum)*, an annual species introduced around 1890, is the dominant grass in much of the Palouse region and southward (page 178). Also present is sagebrush *(Artemisia tridentata)*, a shrub that has increased in importance since human settlement. It is now dominant in much of the southern portion of the region, where open Palouse Prairie formerly bordered the Cold Desert. Except for the presence of shrubs, the Palouse Prairie association resembles the northern Short and Mixed Grass associations.

California Grassland.   This association occupied much of the central valley of California between the Coast Range mountains to the west and the Sierra Nevada to the east. The grassland extended to the woodlands and shrublands of the foothills. California Grassland originally had medium-tall perennial bunch grasses scattered among shorter species (Figure 10.9), thus resembling the Mixed Grass association in appearance. The dominant bunch grass was purple needlegrass *(Stipa pulchra)*, but today the introduced annual grasses

FIGURE 10.9 California Grassland vegetation in central California dominated by purple needlegrass, a bunch grass. *Source.* White, K. L. 1967. Native bunchgrass *(Stipa pulchra)* on Hastings Reservation, California. Ecology 48:949–955. Copyright © 1967 by the Ecological Society of America.

predominate (page 178); these include wild oats *(Avena fatua),* mouse barley *(Hordeum murinum)*, brome grasses *(Bromus* spp.), and fescues *(Festuca* spp.).

Adaptations

Grassland plants generally have most of their biomass below ground, and thus are characterized by high root–shoot ratios. Part of the subterranean biomass of perennials is made up of reproductive structures such as bulbs, corms, or rhizomes. Their position decreases the likelihood of damage from drought. Grasses have the further adaptation that their growth comes from below the soil surface and not from exposed aerial shoot meristems (growing points; Figure 10.3).

In all grassland plants the majority of the below-ground biomass consists of extensive root systems—an obvious adaptation to low soil-moisture levels (Figure 10.10). Competition between different species for moisture and mineral nutrients is reduced by root zonation. All plants have a significant proportion of roots in the upper soil, but some species are restricted to this zone and others have roots that extend much deeper. The deepest root systems are found in the Tall Grass region. For example, big bluestem commonly develops roots that reach depths of two or more meters. In comparison, little bluestem—a species of both the Tall and Mixed Grass regions—forms a dense mass of roots in the top meter of the soil, with a few roots extending to one and one-half meters. Plants of the Short Grass region have still shallower roots, an adaptation that permits water absorption even following light showers.

Shoot adaptations to dry conditions include the narrow leaves of all grasses and many forbs and the finely divided leaves of some forbs. Such leaves are less susceptible to high heat loads than are broad, undivided leaves, since greater air movement results in increased cooling. Apparently, this is more beneficial than the accompanying detrimental increase in transpiration. Other plants, such as prairie dock *(Silphium terebinthinaceum)*, have a vertical leaf orientation with the flat sides facing east and west (Figure 10.11). This maximizes light absorption during the cooler parts of the day, but minimizes heat loads when the sun is directly overhead.

Hairs on the leaf surfaces of many grassland species may reduce transpiration by reducing air flow, but their chief function possibly is the shading of the leaf surface. Another adaptation involves plants that wilt. In many species the stems are kept upright by internal support tissue; however, the leaves usually curl, reducing surface area exposure and, hence, light absorption and transpiration (Figure 10.12). Most grass species have large, thin-walled "motor cells" along the grooves in their leaves. These cells collapse as water is lost in transpiration. The leaves then fold inward, reducing the exposure of stomates, which in grasses are primarily on the upper (inner) surface. Further adaptations that reduce transpiration are the relatively thick cuticles of many prairie plants and the latex sap of the milkweeds *(Asclepias* spp.) and spurges *(Euphorbia* spp.).

Soil-moisture conditions are improved by a thick mulch of organic matter on the soil surface. This layer prevents some rainfall from infiltrating the soil, but it also reduces evaporation from the soil, retards runoff, and reduces erosion. Many grassland plants have leaf arrangements that cup rainfall into the stem where it flows downward to their root zone.

FIGURE 10.10 Stem bases and exposed roots in the upper 10 cm of soil in a 1.8-m square plot of little bluestem prairie. Reprinted from "Prairie Plants and their Environment: A Fifty-year Study in the Midwest" by J.E. Weaver by permission of University of Nebraska Press. Copyright © 1968 by the University of Nebraska Press.

FIGURE 10.11 Prairie dock has broad leaves with a vertical, east–west orientation.

FIGURE 10.12   The wilted, curled leaves of ox-eye *(Heliopsis helianthoides)*, a grass-land forb.

The growth pattern of grassland plants is well adapted to seasonal mois-ture stress. The general timing of growth in the Palouse and California Grasslands has already been mentioned (page 170). Plant activity in the central grasslands matches summer precipitation. Cool season grasses grow in the spring and fall and thereby avoid late summer droughts. Warm season grasses sprout (or germinate) in the spring and reach peak productivity before soil-moisture stress reaches a high level. The timing of productivity and other aspects of growth varies between individuals of the same species across the grasslands. With the early onset of drought and the cool tempera-tures that characterize the Short Grass region, flowering occurs there before it does further east.

Growth is also influenced by the winter season. Perennial herbs are well adapted since their dormant parts are protected beneath the ground. There is evidence that prior to the onset of dormancy, carbohydrates and mineral nutrients are translocated from dying aerial parts to underground tissues.

Many grassland plants are also adapted to periodic long droughts. For example, perennial herbs have life spans of up to 20 years or longer, are able to continually expand their root systems, and are known to produce smaller aerial shoots during drought years. Grassland species also may flower earlier during drought years and/or rely primarily upon vegetative means for reproduction. Some grassland plants have the ability to become dormant during periods of drought and rapidly recover when conditions change. This includes plants that are capable of rapidly revegetating an area by stolons following the end of a drought period.

Fire and grazing are other important aspects of most grassland environments. Again, perennial herbs are well adapted. Although aerial portions are lost when the plants are burned or grazed, the below-ground parts are able to sprout and replace them. Grasses, with their apical buds remaining below the soil surface, are especially capable in this regard; the buds are below the zone of high temperatures during fires and are out of the reach of most grazers. However, repeated burning or grazing over several years can result in the gradual diminishment of below-ground food reserves and the eventual death of many species. Sod grasses, because of their spreading growth-form (vegetative reproduction), are less vulnerable to overgrazing than are bunch grasses.

Human Impact

The aboriginal Indians periodically burned the grasslands. This and lightning fires limited the growth of woody plants and maintained the grassland-forest boundary east of its present position (page 00). The early pioneer settlers of the boundary region avoided the grasslands because of the lack of building materials, fuel, and water and the presence of large fires and strong winter storms. Also, the settlers considered the grasslands infertile because of the absence of trees. Even after it was learned that the Prairie soil was rich, it remained impossible to break the extensive root systems until John Deere developed the steel plow in 1837–1840.

After the invention of the steel plow, settlement of the Tall Grass region proceeded very rapidly, especially as railroad systems opened eastern markets. Extensive roads and plowed fields virtually eliminated fire as a part of the environment. Consequently, trees of the deciduous forest migrated into this climatically borderline region. The same phenomenon has occurred in the north where quaking aspens *(Populus tremuloides)* of the Boreal forest have extended southward into the Mixed Grass region (page 103). Today,

FIGURE 10.13 A stand of Tall Grass prairie along a railroad in northwest Indiana.

few stands of the Tall Grass association remain (Figure 10.13); the rich Prairie soil supports a giant agricultural industry with corn as the primary crop.

Because the dry climate was not suited for agriculture, the Mixed and Short Grass regions were used by the first settlers for open-range cattle ranching. Homestead settlement began in the middle of the nineteenth century and, although settlers were required to cultivate a portion of their land, the dry environment forced them to use most of it for grazing. However, settlement led to the fencing of the range land with barbed wire. This restricted the movement of cattle and therefore increased the likelihood of overgrazing. More of the region was brought into cultivation when dryland farming techniques were developed in 1885–1890. Overgrazing became a greater problem with increasing numbers of cattle on decreased amounts of range land.

Livestock are selective in what they eat; the more palatable plant species are grazed first and heaviest. These usually are the native climax grasses and forbs, especially the legumes. Overgrazing retards root growth, making plants more vulnerable to drought. It also reduces the food storage of perennials, decreasing their vigor. Unpalatable species are therefore at a competitive advantage and become more widespread. Overgrazing also reduces total plant cover, leaving the soil more open to both wind and water erosion (Figure 10.14).

High wheat prices during World War I led to the cultivation of land that was marginal for such agriculture. This further decreased grazing land and, in 1933, a drought began that was the most severe ever recorded in the central grasslands. For seven years there was little precipitation, high temperatures, low humidities, and frequent hot dry winds. With soil exposed by cultivation, overgrazing, and drought, wind erosion became a severe problem; a single storm could remove 2.5 centimeters of topsoil. Great dust storms darkened the sky. In March 1935, erosion from an area 400 kilometers away resulted in the suspension of 37,000 metric tons of soil per cubic kilometer of air above Wichita, Kansas (see J. E. Weaver and F. W. Albertson, Selected Readings, this chapter). Soil from the central Great Plains was carried at least as far east as Chicago. The "Dust Bowl" and the accompanying economic depression caused the abandonment of thousands of farms.

Damage to native vegetation was great. Plants were affected by abrasion, drought, burial, high temperatures, grasshopper invasions, and further overgrazing. Survival was greatly dependent on root development. Vegetation cover was generally reduced by at least 50 percent and frequently by 90 to 98 percent. Weedy annuals and cacti increased, and Mixed Grass vegetation replaced Tall Grass in a band 175 kilometers wide along the boundary between the two associations. The effects of the drought persisted for years after its termination; recovery is thought to have occurred by 1960.

Today, strip mining poses a major problem in much of the northern Short Grass region. Even when the topsoil is returned to place after mining, the regrowth of grassland vegetation is very slow in this region.

In the Desert Grassland, the major vegetation change has been the increase of shrubs, such as mesquite and acacia, that developed with human

FIGURE 10.14   A portion of a Short Grass region that was plowed for wheat cultivation. The exposed soil is highly susceptible to erosion.

settlement. Many factors may have been involved, but cattle grazing is thought to have been the most important. Livestock grazing in this region, which evolved without large grazers, most likely decreased vegetation cover and thereby lowered soil-moisture levels. This and a climatic trend toward drier conditions resulted in an increase of shrubs, most of which are desert species. Livestock may have aided in the seed dispersal of the shrubs. Fires—which would have reduced, if not prevented, the invasion of shrubs—became less frequent and less intense as grazing decreased fuel loads. Evidence indicates that once mesquite becomes established, it spreads quickly, regardless of grazing intensity.

Livestock grazing is also considered the primary cause of vegetation changes in the Palouse Prairie and California Grassland associations. Sagebrush in the Palouse region is favored by the reduced competition that follows the grazing of palatable grasses. The introduced annual grass, downy chess, has become dominant in the Palouse Prairie, partly because its rapid root elongation at low temperatures allows it to elongate during the winter and use soil moisture before the native grasses develop.

The original perennial bunch grasses of the California Grassland evolved without much pressure from large grazers. Thus, these grasses were nearly lost when intensive grazing was initiated by the Spaniards. Today, 50 to 90 percent of the cover of uncultivated areas is made up of introduced species, the majority of which are annuals.

SUGGESTED READINGS FOR FURTHER STUDY

Coupland, R. T. 1958. The effects of fluctuations in weather upon the grasslands of the Great Plains. Botanical Review 24:273–317.

Daubenmire, R. F. 1942. An ecological study of the vegetation of southeastern Washington and adjacent Idaho. Ecological Monographs 12:53–79.

———. 1968. Ecology of fires in grasslands. Advances in Ecological Research 5:209–266.

Franklin, J. F. and C. T. Dyrness. 1973. Natural vegetation of Oregon and Washington. United States Department of Agriculture. Forest Service Technical Report PNW–8. 417 p.

Hastings, J. R. and R. M. Turner. 1965. The changing mile. University of Arizona Press, Tucson. 317 p.

Heady, H. F. 1977. Valley grassland, p. 491–514. *In* M. G. Barbour and J. Major (eds.). Terrestrial vegetation of California. Wiley, New York.

Humphrey, R. R. 1974. Fire in the deserts and desert grassland of North America, p. 365–400. *In* T. T. Kozlowski and C. E. Ahlgren (eds.). Fire and ecosystems. Academic Press, New York.

Johnston, M. C. 1963. Past and present grasslands of southern Texas and northeastern Mexico. Ecology 44:456–466.

Komarek, E. V., Sr. 1965. Fire ecology—grasslands and man. Proceedings, Tall Timbers Fire Ecology Conference 4:169–220.

Lockeretz, W. 1978. The lessons of the Dust Bowl. American Scientist 66:560–569.

McMillan, C. 1959. The role of ecotypic variation in the distribution of the central grassland of North America. Ecological Monographs 29:285–308.

Nicholson, R. A. and G. K. Hulett. 1969. Remnant grassland vegetation in the central Great Plains of North America. Journal of Ecology 57:599–612.

Smeins, F. E. and D. E. Olsen. 1970. Species composition and production of a native northwestern Minnesota tall grass prairie. American Midland Naturalist 84:398–410.

Transeau, E. N. 1935. The prairie peninsula. Ecology 16:423–437.

Vogl, R. J. 1974. Effects of fire on grasslands, p. 139–194. *In* T. T. Kozlowski and C. E. Ahlgren (eds.). Fire and ecosystems. Academic Press, New York.

Weaver, J. E. 1954. North American prairie. Johnsen, Lincoln, Nebraska. 348 p.

———. 1968. Prairie plants and their environment: A fifty-year study in the Midwest. University of Nebraska Press, Lincoln, 276 p.

Weaver, J. E. and F. W. Albertson. 1956. Grasslands of the Great Plains: Their nature and use. Johnsen, Lincoln, Nebraska. 395 p.

11

DESERT VEGETATION

Deserts are usually thought of as sparsely vegetated areas with cacti as the dominant plants; however, there is great variation in deserts and many are not at all like the common image. For example, the Polar desert is an extreme type of Arctic tundra vegetation (page 84).

This chapter considers only those deserts with high summer temperatures. In North America, these occur in two areas (Figure 11.1): one extends from the northwest United States through most of Nevada and Utah to western Mexico, and the other covers portions of southern New Mexico and Texas and northcentral Mexico. These two desert regions are separated by Desert Grassland and other vegetation along the high plains of the Continental Divide.

The western desert region is divided into the Great Basin, Mojave, and Sonoran Deserts; the eastern desert is the Chihuahuan (geographical names are used since they are well established). The Great Basin Desert is often referred to as the Cold Desert, and the other three form the Warm Desert. The vegetation of these deserts was derived primarily from the Madro-Tertiary Geoflora and secondarily from the Arcto-Tertiary Geoflora (page 62).

ENVIRONMENT

An environmental factor common to all four North American deserts is a very low precipitation–evaporation ratio (Figure 11.2). Annual precipitation is highly variable and is usually not evenly distributed, so seasonality results in prolonged periods without significant precipitation. Furthermore, rainfall frequently comes in heavy cloudbursts, during which much of the water is rapidly lost in surface runoff and sometimes forms flash floods.

(*a*) Mojave Desert vegetation in southern California.

Temperatures also show great variation, but on a regular basis. The generally cloudless skies result in large heat gains and high temperatures during the day and rapid losses and cool temperatures at night. The high daytime temperatures and low humidities produce high rates of evaporation. Winter temperatures in the Great Basin Desert, the Cold Desert, are usually below freezing. Winter frosts occur far less frequently in the Warm Desert.

Another general aspect of desert climate is strong wind; this increases evapotranspiration and may cause sand and dust storms, since the soil is largely exposed.

The Great Soil type of the North American deserts is the Grey (or Desert). Grey soils are poorly developed; there is little observable profile differentiation. The soils have small amounts of organic matter, high concentrations of salts, and many of the chemical properties of the parent material. The surface of the soil may be covered with rocks ("rock pavement"), but more fre-

(*b*) A Sonoran Desert cactus in southern Arizona.

quently it is crusty. This accelerates water runoff and may inhibit seedling growth.

Variation in certain soil factors, especially moisture level and salt concentration, is very important in the distribution of communities. The variation is often related to topography. Numerous small mountain ranges are especially common in the north. They receive high amounts of precipitation. Many of the broad valleys between them have depressions that are not externally drained. There, continuous internal drainage and water evaporation result in the deposition of sediments and salts carried by the water. These areas, called playa lake basins, have fine soil with high salt content; vegetational changes occur along a soil salinity gradient.

Details on the environments of the four deserts are discussed later in this chapter.

(c) A stream wash in the Sonoran Desert of southern Arizona.
(d) A Sonoran Desert cactus.

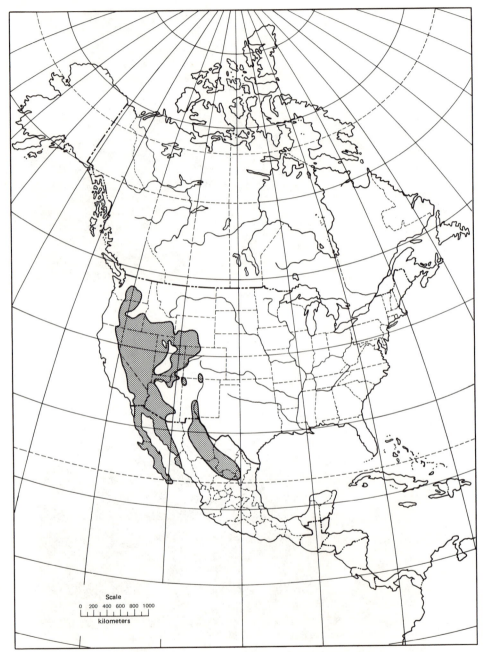

FIGURE 11.1 Distribution map of the deserts.

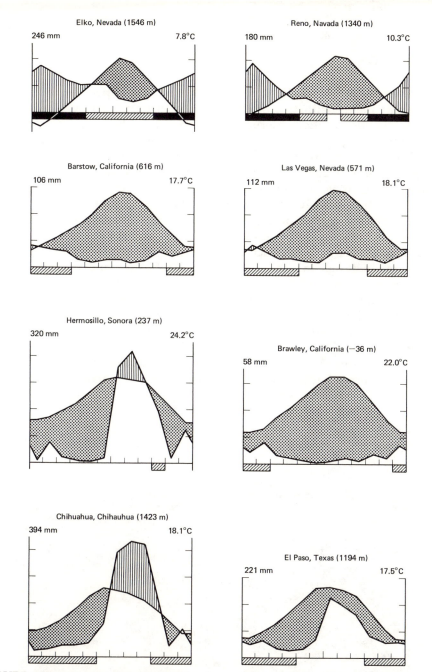

FIGURE 11.2 Climate diagrams for the desert region: Elko and Reno are in the Great Basin (Cold) Desert; Barstow and Las Vegas are in the Mojave Desert; Hermosillo and Brawley are in the Sonoran Desert, but note the difference in precipitation patterns; Chihuahua and El Paso are in the Chihuahuan Desert. *Source.* Redrawn from Walter, H. and H. Lieth. 1967. Klimadiagramm-weltatlas. Fischer, Jena, East Germany.

VEGETATION

Desert vegetation is usually dominated by two low, open shrub layers. In some areas plants with tree or treelike growth-forms may protrude above these layers. Below them there is a discontinuous herbaceous layer. Also important are algae and fungi at the soil surface; they greatly increase soil organic matter (which is still lower than in any other formation) and thereby change the soil structure and increase fertility and infiltration. Nitrogen-fixing blue-green algae are common.

The shrubs are widely spaced (Figure 11.3), and in a few communities appear to have a regular arrangement—an unusual pattern in natural communities. It has been hypothesized that the wide spacing of shrubs is the result of one or both of two factors: the release of chemicals that inhibit the growth of seedlings in the proximity of established individuals and the intense root competition for soil moisture. Evidence indicates that the latter factor is a more likely, yet not a complete explanation of the wide spacing. Many of the herbs are found only beneath the shrubs; this distribution appears to be related to the accumulation of windblown organic matter beneath the shrubs.

Structural diversity and species diversity generally increase southward. Some of the more widespread species and genera are creosote bush *(Larrea divaricata)*, sagebrush *(Artemisia tridentata)*, ocotillo *(Fouqueria splendens)*,

FIGURE 11.3   The wide spacing of shrubs in the Great Basin Desert near Grand Canyon National Park, Arizona.

mesquites (*Prosopis* spp.), and acacias (*Acacia* spp.). There are numerous cases of ecological equivalents in different desert regions. Many times these involve species of the same genus, as in the mesquites.

Seasonal variation is apparent in desert vegetation, despite the persistence of shrubs. Most of the herbs are annuals, and they are abundant following periods of precipitation. However, species composition varies greatly year to year, apparently as a result of the irregular climate. In contrast, perennials, especially the woody species, show little variation over periods of several years; seed germination and seedling survival are generally uncommon.

Typical succession is virtually absent in deserts. Even in highly disturbed areas, the revegetating species are those that will continue to dominate. As in the tundra, the plants apparently do not modify the environment sufficiently to bring about species replacement.

The desert has the lowest annual net productivity of any formation. Values of 70 grams of dry weight per square meter per year are the average. The percent of incoming solar radiation used in gross production is low. The 0.05 percent value given in Table 3.1 is less than one-thirtieth of that for a summergreen (deciduous) forest. Nevertheless, the productivity of individual desert plants is not low and, in fact, may exceed that of some mesophytes. Certain desert plants show yearlong production.

Individual Deserts

Great Basin Desert. This desert, which makes up the Cold Desert, covers the Great Basin physiographic region, extending somewhat further east and west in the north. In this book it is considered to include the Painted Desert of Arizona. The base elevation of the Great Basin is between 1200 and 1600 meters. The topography consists of a series of small north–south oriented mountain ranges of 2500 meters or more, with broad valley floors between them. Many of these valleys are internally drained and hence have playa lake basins.

The Sierra Nevada and Cascade Mountains produce a rain shadow in this region. Annual precipitation values are mostly 10 to 25 centimeters, but can be 7.5 centimeters in the basins and 35 centimeters on the slopes of the mountains (Figure 11.2). The higher elevations of the mountains, especially in the east, receive more precipitation and have woodland and coniferous forest vegetation (Figure 1.2). Precipitation is more evenly distributed throughout the year than in the other deserts, but there is a peak in winter when it falls as snow. Much of the winter precipitation is lost to the dry atmosphere by sublimation, the direct evaporation of snow. Summer precipitation comes in thunderstorms, and thus is irregular and spotty in distribution.

Temperature variations in the Great Basin are typical of a continental climate; the mountains block any moderating influence of the oceans. Temperatures are significantly lower than in other deserts further south. This is the result of the high elevations and the flow of polar continental air from Canada. Winter temperatures seldom rise above freezing, except in the valleys. The frost period extends from early fall until late spring, so the potential growing season is far shorter than in the Warm Desert. Summer frosts are a possibility at night in some localities, including valleys where daytime temperatures may approach 40°C. The combination of relatively moderate temperatures and the even distribution of precipitation results in more favorable soil-moisture conditions than might be expected considering the annual precipitation figures.

Fires are usually not thought of as an environmental factor in most deserts, but they can be important in the denser shrub stands of the Great Basin.

The vegetation is so simple in terms of composition and structure that landscapes often appear quite monotonous, being covered with stands of deciduous or semideciduous greyish-green shrubs. Stand density is low but increases with higher elevations. Woody growth-forms other than shrubs are rare, except above 2000 meters where trees of the Pinyon-Juniper woodland (page 205) are mixed with desert species. Succulents and semisucculents (plants with high proportions of water storage cells, such as cacti) are far less common than in the comparatively frost-free southern deserts. Ephemeral herbs are most prevalent following spring rains. Grasses are important in the north, but less common in the south.

Most of Great Basin Desert vegetation can be divided into two community types whose distribution is based on soil factors, especially soil salinity. One of the communities is dominated by sagebrush and the other by shadscale *(Atriplex confertifolia)*. The other shrub species present are not especially common. Regardless, many times different species of the same genera occur in both communities. These species may appear in a mosaic of single species-dominated stands; examples include winter fat *(Eurotia lanata)*, greasewood *(Sarcobatus vermiculatus)*, and Nuttall saltbush *(Atriplex nuttallii)*.

The Sagebrush community predominates both in the north and on cool slopes above basins (Figure 11.3). In comparison to the Shadscale community, the soil of sagebrush areas tends to be better leached, with lower salt concentrations and a deeper calcium carbonate layer. Sagebrush has a wide distribution that includes areas outside of the Great Basin Desert (Figure 11.4). It occurs on coarse soils where its relatively high oxygen requirements for root growth are met. The Shadscale community is dominant both in the

FIGURE 11.4   A sagebrush-dominated desert community in southcentral Colorado.

southern Great Basin and at the low, warm elevations of basins. The Mormon pioneers in Utah could identify potential agricultural areas by following the distribution of these two communities.

The poorest soils of the Great Basin are those of the floors of the playa lake basins; their texture is very fine and the salt content very high. The vegetation around the playa lakes forms a series of circular belts or zones whose distribution is correlated with soil-salt concentrations.

Mojave Desert.   This is one of the three Warm Desert types. It is the smallest of the North American deserts, covering the southern tip of Nevada and an adjacent part of southern California. The Mojave Desert is frequently considered to be a transition between the Great Basin Desert and the Sonoran Desert to the south. The Mojave Desert region has topography that is similar to that of the Great Basin in that numerous small mountain ranges alternate with low wide valleys. The base elevation varies from 1200 meters at the foot of the mountains in the west to near sea level along the Colorado River in the east (Death Valley has a low of 86 meters below sea level). Most of the region is between 600 and 1200 meters. Drainage is primarily internal, so variations in soil texture and salinity are important in species distributions.

Annual precipitation figures are lower than those of the Great Basin (Figure 11.2). Values decrease from around 12.5 centimeters in the west to 5 centimeters in the east. These low values are the product of a rain shadow during the winter and a constant high-pressure system in the summer (page 30). Precipitation is very irregular; extremely little comes during the summer, and there is a late winter maximum.

Winter temperatures in the basins may reach 15°C during the day and drop below freezing at night. Upper slopes of the mountains have much smaller temperature ranges; winter maxima are around 5°C and minima are below freezing. Freezing temperatures are most common in the western half of the region. Cold air flowing downslope commonly produces temperature inversions (warmer air above colder air) in the winter. This environmental factor is of key importance in species distributions in the region. For example, frost susceptible cacti do not occur in the basins, the sites of cold air accumulation, but only on lower slopes above them. The importance of topography is magnified during the winter when precipitation and slope-exposure insolation differences are at a maximum.

Summer temperatures range from 30 to 40°C in the basins and from 18 to 30°C on the mountain summits. Neither cold-air drainage nor slope-exposure differences are as pronounced as during the winter. The frost-free period is longer than in the Great Basin Desert.

The vegetation of the Mojave Desert resembles that of the Great Basin in several ways, including simple composition and structure and species distributions largely controlled by soil texture and salinity. Nevertheless, since the major dominants are primarily of southern distribution, this desert is considered one of the Warm Desert types.

An open-shrub community dominated by creosote bush and bur sage (*Franseria dumosa*) covers three-fourths of the region (Figure 11.5). There is some variation in height and density. Other significant species include sagebrush, both in the north and at the higher elevations, and the Mojave yucca (*Yucca schidigera*) and Joshua tree (*Yucca brevifolia*) in the west (photograph *a*, page 181). Species adapted to saline conditions are found around the playa lake basins. Cacti and other succulents are not common in most areas.

Sonoran Desert.   This is the most varied of the North American deserts. It ranges from southeast California and central Arizona, south through Baja California, and to a latitude of 27° in western mainland Mexico. Topographic

FIGURE 11.5   Mojave Desert vegetation dominated by creosote bush in Death Valley, California.

relief is generally low. Three-fourths of the region consists of level plains and gentle slopes known as bajadas; the rest of the area is dotted with sandy hills and small mountains. The elevation is the lowest of the deserts, ranging from sea level to 900 meters; most of it, however, is below 600 meters. The highest elevations are found in the north and the east. In contrast to the Great Basin and Mojave Deserts, there are few undrained basins, and the distribution of different communities correlates better with soil texture than with soil salinity.

The Sonoran Desert is not the product of a rain shadow, but instead is a subtropical desert maintained by a persistent high-pressure system. Precipitation is highly variable and irregular. Fog-drip is important along the Pacific Coast of Baja, but normal precipitation averages from near 0 in the west to 35 centimeters in the east (Figure 11.2). Precipitation on mountain tops may total as much as 70 centimeters. Precipitation peaks only in the winter in the western half of the desert, and in winter and summer in the eastern half (compare the climate diagrams for Brawley and Hermosillo). Winter precipitation falls fairly gradually, but summers are characterized by cloudbursts followed by great runoff.

The biseasonal precipitation pattern results in shorter rainless periods than in other deserts; however, the drought periods are more critical here because of higher temperatures—this is the hottest of the North American deserts. Summers are extremely warm, with many days over 40°C. Winters are mild, with highs of around 20°C. Frosts are uncommon, although daily variations in temperatures are great; in barren areas during the summer soil-surface temperatures may fluctuate more than 60°C in a 24-hour period.

Of all the North American deserts, the Sonoran Desert has the greatest diversity of communities and growth-forms. Communities are low, open, and simple along the low elevations of the Colorado River and the Gulf of California, but become richer and more diversified as elevation increases. The southern portions have the greatest growth-form diversity. There are evergreen and deciduous shrubs, stem and leaf succulents, semisucculents, groups of herbaceous ephemerals that correspond to each of the two periods of precipitation, and scattered small trees. With such diversity, the Sonoran Desert frequently has been divided into several sections.

In general, the arid lowland regions of the Sonoran Desert are dominated by bur sage and creosote bush. These are the same dominants as those found over most of the Mojave Desert, but the associated species are different. Communities of greater diversity are found at higher elevations. For example, in southern Arizona and northern Sonora, Mexico, the rocky soils of the slopes above the valleys support communities dominated by palo verde *(Cercidium* spp.)—a tree, as well as saguaro *(Cereus giganteus)*—a giant columnar cactus (Figure 11.6). Although these upland soils have less moisture than the finer valley floor soils, their coarse texture results in greater water availability and root uptake. The saguaro is limited by low water availability at low elevations and by frosts at high elevations. Survival of saguaro seedlings is greatly aided by the shade, concealment, and physical protection provided by both "nurse-plants" and rocks.

Chihuahuan Desert.   This desert and the Sonoran Desert are named after the Mexican states where they are centered. The Chihuahuan Desert extends from the lowlands of southern New Mexico and western Texas through the central portion of the northern third of Mexico. Elevations range from less than 900 meters along the Rio Grande River to over 1800 meters further south in the Mexican Plateau physiographic region. Most of the region is a level plain, interrupted by numerous individual mountains or small mountain ranges. There are many internally drained basins with heavy, saline soils. Much of the soil surface is hard; soil pavement is common.

The Chihuahuan Desert is a subtropical desert, but it has a climate that is generally more moderate than that of the Sonoran Desert. Because of the proximity of the Gulf of Mexico, precipitation is a bit higher than in the other deserts; it increases with elevation from 8 to 50 centimeters (Figure 11.2). Three-fourths of this falls during the summer, producing a growing season that is longer and more certain than those that occur in the other deserts. Except in the lowland north, summer temperatures are 5 to 10°C lower than in the Sonoran Desert. Winter temperatures are also lower, and frosts at night are more common.

FIGURE 11.6 Sonoran Desert vegetation with saguaro cacti near Tucson, Arizona.

The growth-form diversity and hence general appearance of the Chihuahuan Desert is intermediate between the Great Basin and the Sonoran Deserts. Dominance is shared by four types of growth-forms: short, small-leaved shrubs, fairly common but inconspicuous stem succulents, tall semi-succulents, and conspicuous leaf succulents (Figure 11.7). Here the small size of the cacti (the inconspicuous stem succulents) is unlike the cacti of the Sonoran Desert but resembles those of the Mojave Desert. The Sonoran Desert also has more trees than the Chihuahuan. Ephemeral herbs are, of course, present during the summer and survive the dry winters in the seed stage.

Dominant low shrubs include tarbush (*Flourensia cernua*), acacias, and two species of the Sonoran Desert, creosote bush and ocotillo. Mesquite is the only other major shrub in both deserts. The dominant semisucculents are the yuccas (*Yucca* spp.) and similar plants. Leaf succulents include the agaves (*Agave* spp.; Figure 11.8).

FIGURE 11.7 Chihuahuan Desert vegetation in southern New Mexico. The tall semisucculent plants are yuccas *(Yucca elata)*.

FIGURE 11.8 Leaf succulents of the agave genus.

Adaptations

Species found in extreme environments such as the desert or the tundra are of special interest because of their many, varied adaptations. Many of the following adaptations are not necessarily unique to plants of arid regions, but these adaptations are most highly developed in the desert.

Water balance is obviously critical to desert plants, and heat loads are an important related factor. With exposure to high temperatures and bright sunny days, desert plants must be able to tolerate or modify heat loads. It has been shown that some desert species are able to metabolize at very high temperatures. N. F. Hadley reports that many desert plants can survive temperatures of 50 to 55°C, partly because their enzymes and other molecules are highly stable.

Adaptations that reduce heat gain and/or increase loss are common. Vertical leaf or stem orientation is an effective means of limiting the buildup of heat during midday, as is the case for some grassland species. Surface projections such as spines block light, thus reducing heat gain (Figure 11.9; spines also protect plants from some herbivores). N. F. Hadley provides a report of dense spines maintaining surface temperatures of 13°C less than those of unshaded areas. Hairs, especially on leaf surfaces, may perform a similar function. Very dense coverings of spines or hairs also retard heat loss—an advantage in maintaining warm temperatures during the cool nights.

It has been reported that light reflectance is greater in desert plants than in the species of most other regions. The fluted trunks of many columnar cacti reflect a higher proportion of light than do smooth surfaces (Figure 11.10). As these cacti use stored water during dry periods, their succulent stems shrink in accordionlike fashion and increase the depth of fluting. The rough texture and clear character of the cuticle of leaf and stem surfaces are also effective in scattering incoming light, as is the pale coloration of many desert species.

Thick cuticles may also be important in retarding heat loss; for example, most cold resistant cacti have thick cuticular and epidermal layers. In general, succulent stems have a low rate of heat transfer; heat is gained slowly during the day and is retained internally until after sunset when (if the stomates are closed) most heat is lost by reradiation rather than by transpiration. Warmer internal temperatures raise metabolic rates during the night.

Many species have small leaves that aid in preventing the buildup of excessive heat loads. The small leaf size results in great exposure of surface area and, consequently, rapid heat loss. As with grassland species, the benefits of heat loss apparently more than compensate for the negative impact of increased water loss by transpiration.

FIGURE 11.9 The dense covering of spines on this cactus provides protection from light and herbivores.

Many of the above adaptations are also directly important in reducing water loss; however, the adaptive significance of many xeromorphic characters has too often been based on conjecture rather than on evidence. Also, it is important to note that some desert plants transpire at a rate equal to or even greater than that of typical mesophytes. From the perspective of the community, the sparseness and low stature of desert vegetation results in a ratio of water supply per unit of transpiring surface area that approximately equals that of communities in regions with more precipitation.

The chief route of water loss by plants is through the stomates that are open during gas exchange. Many desert species have high resistance to such water loss, since their stomates are sunken in cavities that frequently have numerous hairs or scales. Also, the stomates may be closed under water stress, even during daylight hours. In fact, in many succulents the typical stomatal pattern is reversed; stomates are closed during the day and open at night when cooler air temperatures and lower heat loads result in low transpiration rates. These plants store carbon dioxide as an acid during the night and regenerate it during the day for use in photosynthesis.

Transpiration also takes place through the cuticle, but this layer is very thick in many desert plants. The coating of one Chihuahuan Desert species is so dense that it is collected for commercial use as a sealing wax.

Exposure of transpiring surfaces is perhaps the key factor in water loss. The stems of cacti replace leaves as the photosynthetic organ; therefore, surface area is quite small. The short stature of most desert plants is also adaptive. Furthermore, many desert species are wholly or partly deciduous, losing leaves during dry seasons and producing foliage when soil moisture is available. Creosote bush is an example of a facultative evergreen species; it may lose 80 percent of its leaves during drought, but otherwise retains them. Brittlebush *(Encelia farinosa)* produces leaves of a variety of sizes and morphologies after a period of precipitation. When soil moisture decreases, the less xeromorphic leaves are dropped. Ocotillo, however, sheds all of its leaves during dry periods; new leaves are produced within a few days following precipitation (Figure 11.11). This pattern can occur four to six times a year. Lastly, very dry years may kill whole branches on many desert shrubs and thereby reduce transpiring surface area.

Other species, though they may have some of the above adaptations, are primarily adapted to a desert environment by being drought tolerant, that is, able to survive great degrees of dehydration. These species apparently

FIGURE 11.10 The fluted trunk on saguaro reflects more light than a smooth trunk and allows for expansion when water is stored during periods of precipitation.

FIGURE 11.11   Branches of ocotillo with and without leaves. Leaves form and fall several times in a single year, depending on water availability. An entire multi-branched shrub is shown in the center of Figure 11.6.

undergo a type of "drought hardening" during which their protoplasm becomes more viscous and the remaining water is very tightly bound. Creosote bush, despite its ability to shed leaves during severe drought, is generally not thought to be well adapted to xeric conditions, except for its ability to tolerate desiccation. Another tolerant species is the "resurrection plant" *(Selaginella lepidophylla),* a desert spike moss that is sold commercially because it remains curled and brown when dry, but unrolls and becomes green when wet.

Clearly, the water balance of desert plants also involves at least occasional water uptake. Some species have very deep root systems that may provide aerial parts of the plant with continuous moisture. Mesquite and acacia shrubs reportedly have roots 15 meters or more in length, and bur sage produces roots 5 to 15 times the length of its stem. The term phreatophyte is

FIGURE 11.11   (*Continued*)

applied to those species with roots that extend to the water table. In contrast, other species have shallow root systems; these can effectively use soil moisture from even light rains. Such root systems are often very extensive. For example, the roots of the saguaro cactus are usually less than 1 meter deep, but they may extend horizontally to 30 meters. Many cacti have a shallow, dense network of roots with a few that are deep. In addition, some cacti can produce new roots within 24 hours after a rain. Sagebrush may have some roots extending 3 meters deep, but nearly two-thirds of the root length is typically within 60 centimeters of the soil surface. In spite of such extensive root systems, desert plants, on the whole, do not have high root–shoot ratios, as was the case for grassland species.

Another method of obtaining water is by direct absorption through aerial parts. The hairs on leaves may aid in collecting and retaining fog droplets and dew; however, this moisture must pass through the cuticle to enter the plant. The permeability of the cuticle is increased by the wetting of the leaf surface; nevertheless, the rate of water uptake is slow. Consequently, the chief advantage of surface moisture may be that it lowers rates of transpiration.

Another source of water may be metabolic water—that which is released internally by the plant during respiration. N. F. Hadley reports an experiment which showed that when sections of prickly pear cactus were kept in the dark at 28°C for almost five months, they maintained constant water levels, despite continual transpiration.

Internal moisture storage allows some desert plants, especially succulents, to survive lengthy periods without precipitation. The "barrel" cacti of the Sonoran Desert have become well-known examples of this because of the stories that describe such cacti as a source of emergency drinking water. Old saguaro cacti may weigh six metric tons, and 75 to 95 percent of this weight is water. It has been claimed that such storage allows giant cacti to survive for a year without precipitation.

Another adaptation, drought avoidance, is illustrated by the ephemerals—plants that survive droughts in the seed stage and germinate, grow, and reproduce following precipitation. Deserts with a single wet season have only one set of ephemerals. Areas like the eastern half of the Sonoran Desert have two. The key to the survival of ephemerals is germination only when rainfall is sufficient to permit the completion of their life cycle. Environmental factors important in triggering germination include temperatures following rainfall and the amount and duration of precipitation. Different species undoubtedly respond to various factors; thus, yearly differences in weather result in the great variation of species abundance.

One mechanism of seed germination in ephemerals involves the leaching of growth inhibitors from the seed. A light rain of less than 1.5 centimeters does not completely remove the inhibitor, so no germination occurs. A series of light rainfalls do not have a cumulative effect, since the seed replenishes the chemical inhibitor between rains. Only heavier rainfalls result in germination. Other species may germinate, if there is a sufficient decrease in salinity at the soil surface immediately following precipitation. An operative factor in some species is delayed germination, which means that germination takes place only when relatively heavy rains cause the soil to remain moist for a period of time. Still, other species require the scarification (abrasion) of the seed coat for germination; this occurs when the seeds are carried in water runoff following heavy precipitation.

The growth of ephemerals usually lasts only a few weeks. Typically, the majority of individuals within a population do not survive to maturity. Of those that do mature, their size is determined by available soil moisture. Under dry conditions mature plants may be only 2.5 to 5.0 centimeters in height with a single flower, but under better conditions individuals of the same species may be 35 to 50 centimeters and form numerous flowers.

Human Impact

In general, the aboriginal Indians had little effect on desert vegetation, except in small, localized areas. The first major impact of the Europeans came as a result of livestock grazing. The desert most affected was the Great Basin Desert, where grazing led to an increase in the cover of sagebrush in both the desert region and adjacent areas, such as the southern Palouse Prairie. Forage, of course, has decreased, and the introduced annual grass, downy chess *(Bromus tectorum)*, has become widespread in disturbed areas, especially those that have been overgrazed (page 178). Livestock grazing has also had an impact on other deserts, although the effects are perhaps not as obvious. Grazing in the Sonoran Desert of southern Arizona is thought to have increased runoff and erosion, forming gullies and lowering water tables. It also has decreased the reproduction of the saguaro cactus and other species. Reproduction of the saguaro has been additionally reduced by rodents whose populations have increased following the killing of predators to protect livestock.

Within the last few decades, some of the desert regions of the United States have greatly increased in population. Cities like Phoenix, Arizona, have developed with the rapid expansion of housing tracts that enter the desert areas. In some cases, land has been surveyed, dirt roads bulldozed, and street signs erected so the "developed" land could be sold at highly inflated prices, frequently to distant buyers who have not seen the property. Population growth has resulted in water problems; water tables have greatly dropped as a result of wells, and water systems have been formed to pipe water from rivers such as the Colorado. Much of this water is used to irrigate agricultural fields. However, since evaporation results in the deposition of dissolved materials, continual irrigation may produce soil-salt concentrations that are beyond the tolerance range of crop plants.

Lastly, damage from off-road travel by motorcycles, dune buggies, and four-wheel drive vehicles is rapidly increasing (Figure 11.12). The passage of even a single vehicle is sufficient to disturb the surface crust of the soil. Although the crust will reform after rain, this may not occur for several months and, during that time, the soil is susceptible to wind erosion. Repeated travel can result in soil compaction to depths of at least one meter; this can retard or prevent plant growth. Direct effects on the vegetation include the destruction of annuals through disturbance of their seeds. Perennial plants—with their small stature, brittle nature, and slow growth rate—are also highly susceptible to damage. One study determined that over

FIGURE 11.12 Damage by recreational vehicles to the Mojave Desert. These photographs were taken in 1968 (top), 1970 (middle), and 1972 (bottom) by the Bureau of Land Management. *Source*. They appeared in Stebbins, R. C. 1974. Off-road vehicles and the fragile desert [parts 1 and 2]. The American Biology Teacher 36:203–208, 220, 294–304.

100,000 creosote bushes were severely harmed during a single motorcycle race in southern California in 1973. Repeated use makes such areas barren; revegetation will take several decades, if it occurs at all.

SUGGESTED READINGS FOR FURTHER STUDY

Barbour, M. G. 1973. Desert dogma reexamined: Root/shoot productivity and plant spacing. American Midland Naturalist 89:41–57.

Beatley, J. C. 1975. Climates and vegetation pattern across the Mojave/Great Basin desert transition of southern Nevada. American Midland Naturalist 93:53–70.

Billings, W. D. 1949. The shadscale vegetation zone of Nevada and eastern California in relation to climate and soils. American Midland Naturalist 42:87–109.

Burk, J. H. 1977. Sonoran desert, p. 869–889. *In* M. G. Barbour and J. Major (eds.). Terrestrial vegetation of California. Wiley, New York.

Gates, D. H., L. A. Stoddart, and C. W. Cook. 1956. Soil as a factor influencing plant distribution on salt-deserts of Utah. Ecological Monographs 26:155–175.

Hadley, N. F. 1973. Desert species and adaptation. American Scientist 60:338–347.

Hastings. J. R. and R. M. Turner. 1965. The changing mile. University of Arizona Press, Tucson. 317 p.

Humphrey, R. R. 1974. Fire in the deserts and desert grasslands of North America, p. 365–400. *In* T. T. Kozlowski and C. E. Ahlgren (eds.). Fire and ecosystems. Academic Press, New York.

Jaeger, E. C. 1957. The North American deserts. Stanford University Press, Stanford, California. 308 p.

Johnson, A. W. 1968. The evolution of desert vegetation in western North America, p. 101–140. *In* G. W. Brown, Jr. (ed.). Desert biology: Special topics on the physical and biological aspects of arid regions. Vol. I. Academic Press, New York.

Logan, R. F. 1968. Causes, climates, and distribution of deserts, p. 21–50. *In* G. W. Brown, Jr. (ed.). Desert biology: Special topics on the physical and biological aspects of arid regions. Vol. I. Academic Press, New York.

Mabry, T. J., J. H. Hunziker, and D. R. DiFeo, Jr. 1977. Creosote bush: Biology and chemistry of *Larrea* in New World deserts. U.S./IBP Synthesis Series No. 6. Dowden, Hutchinson & Ross, Stroudsburg, Pennsylvania. 284 p.

McCleary, J. A. 1968. The biology of desert plants, p. 141–194. *In* G. W. Brown, Jr. (ed.). Desert biology: Special topics on the physical and biological aspects of arid regions. Vol. I. Academic Press, New York.

Orians, G. H. and O. T. Solbrig. 1977. Convergent evolution in warm deserts: An examination of strategies and patterns in deserts of Argentina and the United States. U.S./IBP Synthesis Series No. 3. Dowden, Hutchinson & Ross, Stroudsburg, Pennsylvania. 333 p.

Shreve, F. 1942. The desert vegetation of North America. Botanical Review 8:195–246.

Shreve, F. and I. L. Wiggins. 1964. Vegetation and flora of the Sonoran Desert. Vol. I. Stanford University Press, Stanford, California. 186 p.

Stebbins, R. C. 1974. Off-road vehicles and the fragile desert [parts 1 and 2]. The American Biology Teacher 36:203–208, 220, 294–304.

Steenbergh, W. F. and C. H. Lowe, 1969. Critical factors during the first years of life of the saguaro *(Cereus giganteus)* at Saguaro National Monument, Arizona. Ecology 50:825–834.

Vasek, F. C. and M. G. Barbour. 1977. Mojave desert scrub vegetation, p. 835–867. *In* M. G. Barbour and J. Major (eds.). Terrestrial vegetation of California. Wiley, New York.

Walter, H. and E. Stadelmann. 1974. A new approach to the water relations of desert plants, p. 213–310. *In* G. W. Brown, Jr. (ed.). Desert Biology: Special topics on the physical and biological aspects of arid regions. Vol. II. Academic Press, New York.

Went, F. W. 1949. Ecology of desert plants. II. The effect of rain and temperature on germination and growth. Ecology 30:1–13.

West, N. E. and K. I. Ibrahim. 1968. Soil-vegetation relationships in the shadscale zone of southeastern Utah. Ecology 49:445–456.

Yeaton, R. I., J. Travis, and E. Gilinsky. 1977. Competition and spacing in plant communities: The Arizona upland association. Journal of Ecology 65:587–595.

Young, J. A., R. A. Evans, and J. Major. 1977. Sagebrush steppe, p. 763–796. *In* M. G. Barbour and J. Major (eds.). Terrestrial vegetation of California. Wiley, New York.

12

TEMPERATE SHRUBLAND, WOODLAND, AND SAVANNA VEGETATION

These three formations, though placed together in a single chapter, have few similarities in structure and composition. Shrubland (thicket) vegetation is dominated by a dense layer of shrubs and/or small, shrubby trees. Woodland vegetation is dominated by trees but, in contrast to forests, most of the crowns do not touch. Savanna vegetation is even more open; trees (or shrubs) have a cover of less than 30 percent, and the dense herbaceous layer is the best developed stratum.

This chapter will consider the woodlands and shrublands of the Rocky Mountain region, the same vegetation and the related forest type of the California area, and, lastly, the savanna of the central United States (Figure 12.1).

THE ROCKY MOUNTAIN REGION

Pinyon-Juniper Woodland

The Pinyon-Juniper woodland is best developed in the Great Basin and southern Rocky Mountains (Figure 12.2). In the west it reaches the east slope of the Sierra Nevada Mountains (often as pinyon or juniper woodland; photograph *a*, page 206) and extends from Arizona and New Mexico to southern Idaho, with variations north to southern Canada and south to northern and central Mexico. Along the east slope of the Rocky Mountains,

(a) Pinyon-Juniper woodland vegetation in western Nevada.
(b) Scrub Oak-Mountain Mahogany shrubland vegetation in southwest Colorado.

(c) A manzanita shrub in the California Chaparral in Sequoia National Park, California.

(d) An oak savanna in eastcentral Minnesota.

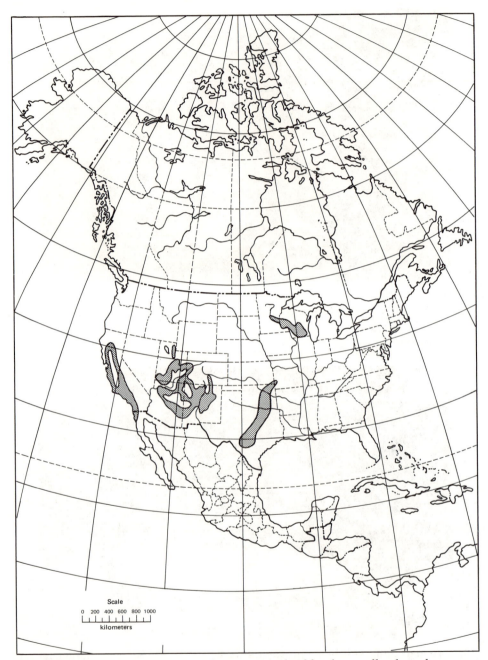

FIGURE 12.1 Distribution map of temperate shrubland, woodland, and savanna vegetation.

FIGURE 12.2   Pinyon-Juniper woodland vegetation dominated by oneseed juniper *(Juniperus monosperma)* in central New Mexico.

true Pinyon-Juniper vegetation is found from northeastern Mexico to central Colorado, but a variation extends further northward. Not all of these areas can be shown on a map of the scale of Figure 12.1.

Pinyon-Juniper vegetation has a discontinuous distribution along high mesas, the foothill zone of large mountains, and the entire elevation of small mountains. Its elevational range varies with location, but in most areas it is between 1500 and 2250 meters. Ponderosa pine forest may be upslope, and desert or grassland vegetation is downslope. In total, Pinyon-Juniper vegetation covers about 190 million hectares in the United States.

The climate of the Pinyon-Juniper zone is intermediate between that of the lower montane (ponderosa pine) zone and that of desert and Short Grass regions (Figure 12.3). Annual precipitation is usually between 30 and 40 centimeters. Summer precipitation falls in cloudbursts, which result in a high percentage of runoff. Little moisture infiltration occurs beyond 30 centimeters; an average depth is 15 centimeters. Mean summer temperatures are about 20°C, with highs averaging near 30°C. Potential evaporation is greater than precipitation. The growing season ranges from 90 to just over 200 days, depending on location.

Soil factors, especially physical ones, are important in the distribution of individual stands; Pinyon-Juniper communities typically occur on coarse soils, such as those found on ridges and steep slopes. These soils have greater available soil moisture. The fine soils of mesas, valleys, and gentle

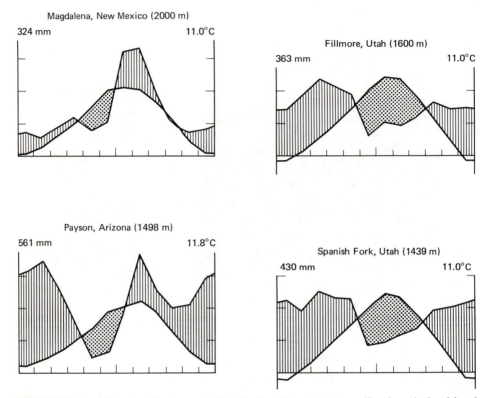

FIGURE 12.3 Climate diagrams in the Rocky Mountain woodland and shrubland region: Magdalena and Fillmore are in regions of Pinyon-Juniper woodland vegetation; Payson and Spanish Fork are in regions in or near Scrub Oak-Mountain Mahogany shrubland vegetation. *Source.* Redrawn from Walter, H. and H. Lieth. 1967. Klimadiagramm-weltatlas. Fischer, Jena, East Germany.

slopes more often support desert or grassland vegetation. In addition, topographic relief is important on both local and regional scales, as already mentioned. Fires were frequent in the past.

The overall appearance of Pinyon-Juniper vegetation is much the same throughout its range. There are three strata: herb, shrub, and tree. On the more mesic sites, the tree canopy may be forestlike in cover, but elsewhere the trees are short (3 to 7 meters) and widely spaced (canopy coverage averages around 50 percent). Individuals are often as broad as they are tall.

There is great regional variation in stand composition. The junipers include *Juniperus occidentalis, J. scopulorum, J. monosperma,* and *J. osteosperma.* There are fewer species of pinyon pines, and as a group they have a higher elevational range; common species include *Pinus edulis, P. mono-*

phylla, and *P. cembroides.* Most shrubs and herbs of Pinyon-Juniper stands are species that also occur at lower elevations (consequently, they are well adapted to dry conditions). Some shrub species are dominants of the Scrub Oak-Mountain Mahogany shrubland (see next section). Grasses dominate the herb layer.

The foliage of conifers has already been discussed as adaptive to xeric environments (page 105). Junipers have appressed scalelike leaves—foliage that is more xeromorphic than pine needles (partially explaining the elevational differences in their distributions). The trees also have widely spreading, dense root systems.

Human impact on the Pinyon-Juniper woodland has occurred from aboriginal times to the present. Although the Indians used pinyon pine seeds as an important wild food, this probably had little effect on the vegetation. More importantly, they and the Europeans used the Pinyon-Juniper zone as a source of fuel wood and timber. Juniper timbers were used in Indian pueblo construction and later for fence posts. Pinyon pine timbers were used for supports in mines.

More widespread impact has resulted from livestock grazing. Much of the region has been overgrazed and this has affected tree regeneration both inside and outside stands. In many areas the woodland has increased in density and invaded nearby grasslands. It has been hypothesized that prior to extensive grazing, dense grass cover competed with conifer seedlings, especially junipers, slowing their rate of growth. The grass cover also fueled fires that periodically killed seedlings as well as larger trees. Most areas of juniper invasion are sites where woodland vegetation had been burned off. Shrub cover may increase following burning and/or grazing, especially in the Pinyon-Juniper zone of the Great Basin region where sagebrush (*Artemisia tridentata*) has become widespread. Relatively recently, mechanized equipment has been used to clear sizable areas of the woodland and to seed grasses in order to increase livestock forage.

Scrub Oak-Mountain Mahogany Shrubland

Shrublands are widespread in the Rocky Mountain region, but in many sites they are either seral postfire communities or have only a local distribution on talus slopes, along streams, and around lakes.

The Scrub Oak-Mountain Mahogany type of shrubland is considered a climax community (Figure 12.4). It ranges from southern Wyoming and northern Utah to the mountains of Mexico and occurs as a band between coniferous forest and desert vegetation, usually at 1250 to 1750 meters, but higher in the central Rockies (where it may be above a Pinyon-Juniper zone). It has much the same precipitation but somewhat warmer temperatures than

FIGURE 12.4 Scrub Oak-Mountain Mahogany shrubland vegetation dominated by scrub oak in central Arizona.

the Pinyon-Juniper woodland. Relief and soil factors seem to be important in the distribution of individual stands.

Similarities in the environments relate Scrub Oak-Mountain Mahogany vegetation more to the coniferous forest and woodland than to the desert. The appearance of the vegetation can vary from dense thickets to clumps with scattered open areas of desert or grassland vegetation. The greatest shrub cover occurs on steep slopes. Fires stimulate reproduction, since some species stump sprout, and stand density decreases with maturity.

The species composition of the vegetation is also quite variable. The greatest floristic diversity occurs in the southern Rocky Mountains. Here, scrub oak *(Quercus turbinella)* and evergreen shrubs dominate. Some species of the California shrublands and woodlands are present. This reflects a common origin of the vegetation of the two regions. The Madro-Tertiary Geoflora was originally centered in Mexico; some of its species migrated and adapted to California's Mediterranean climate and some to the Rocky Mountain environment.

To the north, Gambel's oak *(Quercus gambelii)* is predominate from the southern Rocky Mountains to central Colorado (photograph *b,* page 206), and mountain mahogany *(Cercocarpus* spp.) is common from Colorado to southern Wyoming.

The Scrub Oak-Mountain Mahogany shrubland has been affected by sheep grazing, but there are conflicting reports on whether it has increased or decreased shrub cover.

THE CALIFORNIA REGION

Chaparral

Early Spanish explorers referred to the shrubland vegetation of California as *chaparro*—a term that originated from *chabarra,* a Basque word for a scrub oak of the Pyrenees Mountains. Similar vegetation occurs only in a few other small areas of the world: the Mediterranean region, central Chile, southern Australia, and southwest coastal Africa. In North America, the term chaparral is sometimes applied to the Scrub Oak-Mountain Mahogany vegetation of the Rocky Mountains; however, in this book it is reserved for the shrubland vegetation that ranges from southern Oregon to northern Baja California. Foothill Chaparral is common on steep, low elevation slopes (Figure 12.5); stands of the less widespread Montane Chaparral are upslope, scattered through the coniferous forest zones of the taller mountains.

The California Chaparral region has the cool, wet winters and hot, dry summers of the Mediterranean climate (Figure 12.6). Annual precipitation usually ranges from 35 to 75 centimeters (with the lower amounts in the south). Even greater precipitation may occur in the north and at the higher elevations. At least 80 to 90 percent of the precipitation falls from November to April. During the winter, mean minimum temperatures are 0 to 10°C; summer mean maxima reach 30 to 35°C.

The moderate to high temperatures combined with low amounts of precipitation result in significantly less soil moisture where there is Foothill Chaparral than in the higher elevation coniferous forest zones. The timing of plant activity is largely determined by the interaction of soil moisture and temperature. Soil moisture is greatest from November to July. Depending on the species and location, the growth period ranges from 8 to 12 months.

Topography is an important environmental variable; the typically very steep relief enhances runoff and erosion. Slope exposure has a large influence

FIGURE 12.5 Foothill Chaparral dominated by ceanothus in Sequoia National Park, California. Higher elevation coniferous forest vegetation is visible on the ridge at the upper left.

on the distribution of chaparral species. The soils are generally poorly developed and therefore are shallow and coarse, if not rocky. Surface fertility and water-holding capacity are low. Parent material is commonly fractured to great depths.

Summer fires are so frequent that Chaparral vegetation is considered a "fire-type"; that is, its general character is maintained by occasional but periodic burning.

California Chaparral vegetation is dominated by a 1- to 3-meter tall shrub layer that frequently is so dense that it is virtually impenetrable. A few scattered trees may be present. The herb layer is very sparse. Nearly all of the shrubs have the broad-sclerophyll growth-form; that is, their leaves are flat, thick, and evergreen with a heavy cuticle.

The most common species of the California Chaparral, however, is cham-

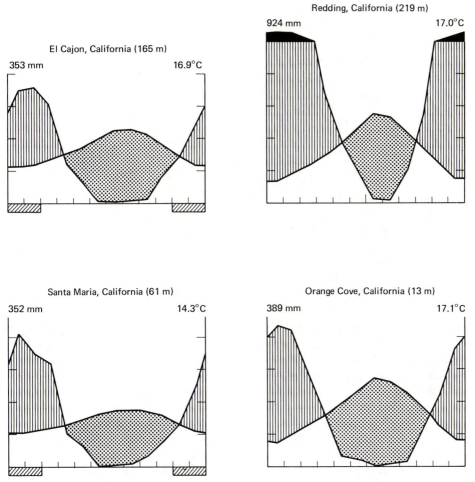

FIGURE 12.6   Climate diagrams in the California Foothill Chaparral, Woodland, and Broad-Sclerophyll Forest region: El Cajon and Redding are southern and northern locations in the Chaparral and Forest region. Santa Maria and Orange Cove are in the California Woodland region. *Source.* Redrawn from Walter, H. and H. Lieth. 1967. Klimadiagramm-weltatlas. Fischer, Jena, East Germany.

ise *(Adenostoma fasciculatum)*, a species with narrow leaves resembling those of a spruce or fir. Chamise is especially predominant in southern California and at low elevations; its stands typically have only a few minor associated shrub species. Other stands of Chaparral are more mixed; there are around 50 species of evergreen shrubs in the California Chaparral. Aside from chamise, the most common are the manzanitas *(Arctostaphylos* spp.; photograph *c,* page 207), which are more widespread at high elevations, and various species of ceanothus (*Ceanothus* spp.), which are more important in the northern half of the Chaparral region. Ceanothus shrubs may have nitrogen-fixing bacteria in root nodules.

The distribution of different California Chaparral communities is influenced primarily by fire (and secondarily by slope exposure and elevation). In the first growing season after a fire, herb and shrub seedlings and sprouts appear in very large numbers (Figure 12.7). Five to ten years later, the herbs decline to where they are nearly absent, apparently as a result of chemical inhibitors released by some of the shrubs. Within 15 years the stand has much the same composition and structure as it did before the fire. A mature stage is reached when long-lived species attain dominance, roughly 30 years. If fires do not occur every 15 to 40 years, short-lived shrubs such as ceanothus die out and stands develop a high proportion of dead wood, have little annual growth, and lack seedlings. It has been hypothesized that the toxic compounds released by shrubs may cause this senescence.

California Chaparral species are well adapted to fire. One of the more common adaptations is the ability of many shrub species to sprout from latent, below-ground buds. This explains why community composition does not change much following a fire. Another adaptation is that many species produce seeds that remain dormant for long periods of time, but germinate following fire. Some herbs have seeds that are inhibited by toxins released by the shrubs, but fires remove the toxins. The seeds of some shrub and herb species may be heat resistant or survive fires while buried in the soil. Some species may also produce seeds whose germination is stimulated by exposure to heat. These adaptations result in great numbers of sprouts and seedlings following fire; H. H. Biswell gives a figure of as high as 3000 per square meter. Several other factors also speed revegetation; these include vegetative reproduction by branch layering and rhizomes sprouting, as well as sexual reproduction just three to five years after a fire.

It has been proposed that species of vegetation that is fire-dependent, such as the California Chaparral, are further adapted to fire in that they have qualities that increase the probability of fire. Chamise is an example of this type; it forms dead branches after reaching an age of 15, has a high surface area-to-volume ratio (as a result of thin stems and small leaves), and, like most Chaparral species, has resinous, highly flammable leaves. Furthermore, its litter is porous and highly susceptible to fire. It has been said that chamise stands depend on fire every 15 to 20 years for optimum vigor and stability.

Adaptations to climate include a general growth pattern of activity that begins after fall rains, ceases with cool winter temperatures, resumes in the spring, and continues until soil moisture is no longer available in midsummer. Summer drought has resulted in adaptations resembling those discussed for desert plants. For example, many shrub species have deep root systems that may penetrate the fractured bedrock to depths of around 8

FIGURE 12.7 (*a*) A photograph taken shortly after a fire. (*b*) and (*c*) Two others taken during the following two growing seasons in the southern California Foothill Chaparral. *Source.* Vogl, R. J. and P. K. Schorr. 1972. Fire and manzanita chaparral in the San Jacinto Mountains, California. Ecology 53:1179–1188. Copyright © 1972 by the Ecological Society of America.

meters; others have shallow, wide-spreading roots. The thick cuticle on aerial parts reduces water loss and reflects some of the incoming light, as does light leaf coloration. Transpiration is also reduced by partial closure of stomates during times of moisture stress. Some species have a vertical leaf orientation. Lastly, many shrub species, including those that are evergreen, may drop some leaves during summer drought; chamise is a good example of this. The evergreen habit, however, allows for rapid resumption of growth following rainfall.

The disturbance of coniferous forests by burning, mining, and logging has led to the expansion of Chaparral vegetation. Many northern California stands are the result of these activities. Overgrazing of low elevation woodlands has led to an increase of Chaparral shrubs there, too. In addition, a change in fire frequency has had a major impact throughout most of the Chaparral region. Fire suppression has led to the development of old age stands; these have high fire hazards and little forage for wildlife populations. In some areas, prescribed burning recently has been used to reduce these problems. Such burning has also been employed to convert areas of Chaparral to grasslands; repeated fires are used to kill off the shrubs, and forage grasses are seeded on the site. An additional impact has been the spread of suburban housing developments into Chaparral areas of southern California.

Broad-Sclerophyll Forest

This forest vegetation has much the same range as California Chaparral. As in all forests, the tree layer is so dense that the crowns interlock (Figure 12.8). Individual trees are 5 to 10 meters in height. An open shrub layer, largely of Chaparral species, and a herbaceous layer are also present. The forest occurs on more mesic sites than Chaparral; hence, it is more common than Chaparral in the north, but less common in the south. In elevation, the forest may extend up to the lower montane coniferous forest zone at approximately 1500 meters. Mean annual precipitation values increase from 35 to 100 centimeters northward. Temperatures seldom drop much below freezing, and in summer they commonly reach over 30°C. Plant growth is limited by cool temperatures in winter and by low soil-moisture levels in mid to late summer.

There is floristic variation throughout the range of this forest; some of the common species include canyon live oak *(Quercus chrysolepis)*, coast live oak *(Quercus agrifolia)*, interior live oak *(Quercus wislizenii)*, madrone *(Arbutus menziesii)*, digger pine *(Pinus sabiniana)*, coulter pine *(Pinus coulteri)*, and bay *(Umbellularia californica)*. All of these species have the broad-sclerophyll growth-form.

As with the Chaparral species, the various leaf characters are thought to

FIGURE 12.8   California Broad-Sclerophyll Forest vegetation dominated by canyon live oak in Sequoia National Park, California. The ridge in the background is in the lower montane zone of the coniferous forest.

be adaptive to summer drought. A different adaptation is illustrated by California buckeye *(Aesculus californica)*. It is deciduous, losing all its leaves when soil-moisture levels become limiting.

Humans have had little impact on this forest.

Woodland

California Woodland vegetation generally occurs at low elevations and is especially common in the interior valleys of California from near Los Angeles to the northern part of the state. Its usual elevational range is from 150 to 1000 meters. Its climate is similar to that of the Broad-Sclerophyll forest, but its soils are generally deeper and better developed than in either forest or Chaparral stands.

The California Woodland has an open canopy (most crowns do not touch) and a dense herbaceous layer (Figure 12.9). The trees are usually 5 to 10 meters in height; their cover values average around 50 percent, but in some areas they are as low as a savanna (less than 30 percent). The herbaceous layer is dominated by plants of the California Grassland association (page 170). Shrubs are uncommon in undisturbed stands. Dominant tree species are blue oak *(Quercus douglasii)*, California white oak *(Quercus lobata)*, and

FIGURE 12.9   California Woodland vegetation dominated by blue oak near Sequoia National Park, California. The denser California Chaparral vegetation covers the foothills in the background.

coast live oak. The latter two have such extensive root systems that they reach the water table. Blue oak is drought deciduous and is generally found on dry upper slopes.

The primary human impact on the California Woodland has come through livestock grazing. Not only has this resulted in a changed herbaceous layer (page 178), but by decreasing competition for oak seedlings it has also increased tree densities. The trees of some lowland stands have been cut for a variety of reasons, including increased livestock forage.

THE GRASSLAND-DECIDUOUS FOREST BOUNDARY REGION

Some scientists restrict the use of the term savanna to tropical vegetation with a dense, grass-dominated herbaceous layer and a tree layer whose canopy coverage is less than 30 percent. Others also apply the term to temperate vegetation with a similar appearance. Using this interpretation, temperate North American examples include the open pinewoods of the southeast United States, parts of the original forests of the central valley of Oregon, the more open stands of the Woodland vegetation of California, and parts of the grassland-deciduous forest transition area of the central United States (Figure 12.10 and photograph *d*, page 207). The following discussion will deal with this last region.

At the time of pioneer settlement, much of the transition area consisted of a mosaic of grassland and forest stands, especially in the prairie peninsula region. D. P. White published evidence that the forest stands occurred on soils where mycorrhizal relationships developed, and grassland vegetation dominated on the other soils. Savanna vegetation was more common to the north and south of this region; in Minnesota, Wisconsin, Oklahoma, and Texas it formed a belt of 75 to 175 kilometers or more in width between the Tall Grass prairie and the deciduous forest (Figure 12.1). In addition, many isolated stands of savanna extended into both the grassland and forest regions.

The environment of the savanna belt is generally intermediate between that of the deciduous forest and that of the grassland (Figure 12.11). Soil is a key factor in local distributions, especially in the south where fine, heavy soils generally support grassland vegetation and coarser, lighter soils have savanna stands. Fire is of great importance; in fact, the open nature of the savanna stands is dependent on fire.

The herbaceous layer of the savanna consists of grassland species, and the trees are Oak-Hickory association dominants. Bur oak *(Quercus macrocarpa)* is the most common tree in the north; post oak *(Quercus stellata)* and blackjack oak *(Quercus marilandica)* dominate the "Cross-Timbers" region of Oklahoma and Texas. Further south in Texas, junipers and mesquite *(Prosopis* spp.) are common associates. Hickories *(Carya* spp.) are common only in stands near the forest region. The shapes of savanna and forest trees, even of the same species, are quite different from one another. Savanna trees have a

FIGURE 12.10   Savanna vegetation dominated by bur oak in western Minnesota.

highly branched trunk with spreading limbs; forest trees are relatively narrow with few low branches.

The root systems of savanna trees are adapted to the dry climate. J. E. Weaver reported that the excavation of mature bur oaks has exposed tap roots 4 meters deep with many branches. One tree had 30 main branches within the top 60 centimeters, and these spread 6 to 18 meters away from the trunk before turning downward; some of the branches may reach greater depths than the tap root. Additional studies have indicated that bur oak seedlings have greater root growth than some mesic species.

Adaptations to fire are essential to the survival of savanna plants. Those of grassland species have already been discussed (page 175). Savanna trees develop thick, somewhat fire-resistant bark and are capable of root sprouting. Bur oak forms woody plates near the soil surface, and sprouts grow from these plates after fires (Figure 12.12). If these sprouts are not burned for a period of 12 to 15 years, they can reach a size at which they are capable of surviving fires. Oak species restricted to the forest region do not have such a degree of adaptation to fire.

The savanna of the central United States has greatly changed as a result of human activity. In Wisconsin and other northern areas, oak savanna was gone just 25 to 30 years after settlement. At least part of the savanna on any settler's land was cut and plowed. Uncut savanna stands had great increases in tree densities, as reduced fire frequency resulted in the greater survival of sprouts and the invasion of fire-susceptible species. These former savanna stands became farm woodlots. In the south, the savanna was heavily grazed.

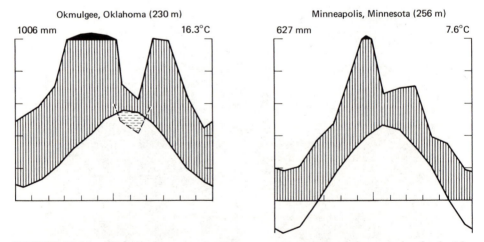

FIGURE 12.11   Climate diagrams in the savanna portion of the grasslands-deciduous forest boundary region. *Source.* Redrawn from Walter, H. and H. Lieth. 1967. Klimadiagramm-weltatlas. Fischer, Jena, East Germany.

FIGURE 12.12   Sprouts of bur oak in a Tall Grass prairie in southern Wisconsin.

This reduced herbaceous cover, leaving the tree seedlings with less competition. Also, fuel loads were lowered, so savanna fires were not frequent enough nor sufficiently intense to kill back seedlings and sprouts.

Today, in both the north and south, the former oak savannas have a woodland or forest appearance. For example, tree density in the Wisconsin savanna region is now ten times what it had been. Even the species composition has changed. Yet, the presence of old, spreading "wolf" trees scattered among the younger, narrow forest trees indicates the previous existence of savanna vegetation.

SUGGESTED READINGS FOR FURTHER STUDY

Biswell, H. H. 1974. Effects of fire on chaparral, p. 321–364. *In* T. T. Kozlowski and C. E. Ahlgren (eds.). Fire and ecosystems. Academic Press, New York.

Bray, J. R. 1960. The composition of savanna vegetation in Wisconsin. Ecology 41:721–732.

Brown, H. E. 1958. Gambel oak in west-central Colorado. Ecology 39:317–327.

Cable, D. R. 1975. Range management in the chaparral type and its ecological basis: The status of our knowledge. USDA Forest Service Research Paper RM-155. 30 p.

Cooper, W. S. 1922. The broad-sclerophyll vegetation of California: An ecological study of chaparral and its related communities. Carnegie Institute of Washington Publication 319. 124 p.

Curtis, J. T. 1959. The vegetation of Wisconsin: An ordination of plant communities. University of Wisconsin Press, Madison. 657 p.

di Castri, F. and H. A. Mooney (eds.). 1973. Mediterranean type ecosystems: Origin and structure. Springer-Verlag, New York. 405 p.

Dyksterhuis, E. J. 1948. The vegetation of the Western Cross Timbers. Ecological Monographs 18:325–376.

Griffin, J. R. 1977. Oak woodland, p. 383–415. *In* M. G. Barbour and J. Major (eds.). Terrestrial vegetation of California. Wiley, New York.

Hanes, T. L. 1977. California chaparral, p. 417–469. *In* M. G. Barbour and J. Major (eds.). Terrestrial vegetation of California. Wiley, New York.

Hayward, C. L. 1948. Biotic communities of the Wasatch chaparral, Utah. Ecological Monographs 18:473–506.

Johnsen, T. N., Jr. 1962. One-seed juniper invasion of northern Arizona grasslands. Ecological Mongraphs 32:187–207.

Merkle, J. 1952. An analysis of a pinyon-juniper community at Grand Canyon, Arizona. Ecology 33:375–384.

Sawyer, J. O., D. A. Thornburgh, and J. R. Griffin. 1977. Mixed evergreen forest, p. 359–381. *In* M. G. Barbour and J. Major (eds.). Terrestrial vegetation of California. Wiley, New York.

St. Andre, G., H. A. Mooney, and R. D. Wright. 1965. The pinyon woodland zone in the White Mountains of California. American Midland Naturalist 73:225–239.

Sweeney, J. R. 1968. Ecology of some "fire type" vegetations in northern California. Proceedings, Tall Timbers Fire Ecology Conference 7:111–125.

Vasek, F. C. and R. F. Thorne. 1977. Transmontane coniferous vegetation, p. 797–832. *In* M. G. Barbour and J. Major (eds.). Terrestrial vegetation of California. Wiley, New York.

Weaver, J. E. 1954. North American prairie. Johnsen, Lincoln, Nebraska. 348 p.

Wells, P. V. 1962. Vegetation in relation to geological substratum and fire in the San Luis Obispo Quadrangle, California. Ecological Monographs 32:79–103.

White, D. P. 1941. Prairie soil as a medium for tree growth. Ecology 22:399–407.

Woodbury, A. M. 1947. Distribution of pigmy conifers in Utah and Northeastern Arizona. Ecology 28:113–126.

13

TROPICAL VEGETATION

North American tropical vegetation occurs from South America to the north through Central America into southern and central Mexico and the islands of the Caribbean—an area nearly the size of the conterminous United States. Subtropical vegetation extends further northward in Mexico and covers the southern tip of Florida (Figure 13.1). The climate throughout both regions is generally warmer and wetter than in the temperate zone.

Most of Central America and many Caribbean islands are ribbed by mountains. Only the Yucatan Peninsula and coastal and valley areas have low relief. The mountains themselves and their interaction with the prevailing winds produce great climatic and vegetational variations. In fact, there are as many different kinds of vegetation in this region as there are in the rest of North America. Consequently, generalizations about the greatest species diversity, stablest climate, longest period without disturbance, and closed-nutrient cycles fit only particular kinds of tropical vegetation.

Classification of tropical vegetation is largely based on dominant growth-forms as related to climate. Detailed classification has been limited by the great species diversity of certain types of vegetation and the relative paucity of floristic ecological research. The type of vegetation that this chapter focuses on is the Lowland Tropical Rain Forest. Detailed consideration of it is followed with discussions of variations along elevational gradients in mountains and seasonality gradients in climate. Last to be considered is tropical savanna vegetation.

LOWLAND TROPICAL RAIN FOREST

Extensive Lowland Tropical Rain Forests are located in three regions of the world: southeast Asia, central Africa, and central and northern South Amer-

(a) Tropical vegetation in the Blue Mountains of Jamaica.

(b) A tropical tree fern.

(c) Tropical vegetation in the Blue Mountains of Jamaica.
(d) Lowland tropical vegetation in northeast Jamaica.

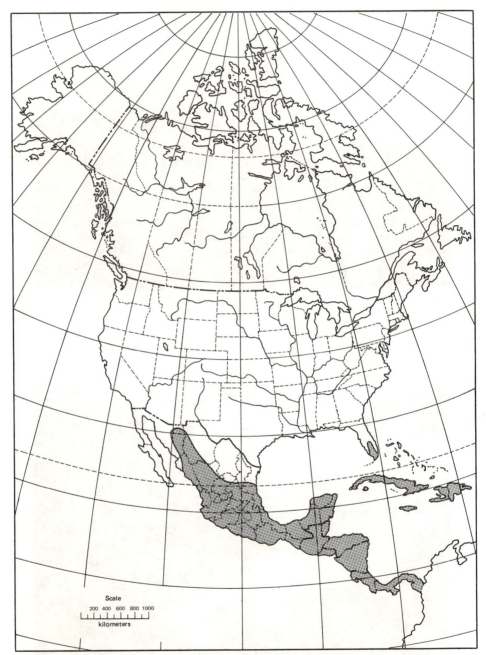

FIGURE 13.1 Distribution map of tropical and subtropical vegetation.

ica. From this last area, rain forest vegetation extends north along the east coast lowlands as far as the Yucatan Peninsula of southern Mexico.

This vegetation was derived from the Neotropical Tertiary Geoflora. It retreated southward from its more northern distribution in the Tertiary Period as a result of the general cooling and drying trends discussed in Chapter 6. It is thought to have changed little in overall appearance since that time.

Environment

The chief characteristic of the Lowland Tropical Rain Forest environment is its consistency. This is best illustrated by the climate, which lacks the seasonality of other types of tropical and temperate vegetation (Figure 13.2). Daily variations in climate are, in fact, greater than annual variations. The mean annual temperature is between 22 and 27°C. Monthly means vary only two to four degrees from the annual mean; the daily range of temperatures somewhat exceeds this.

Precipitation is usually very high on an annual basis but can be quite variable from location to location, and even from month to month. Nevertheless, it is always in excess of evaporation. True rain forest vegetation is not found where there are prolonged droughts; however, short periods without rainfall may be compensated for by prior periods of high precipitation. Relative humidity is always at or near saturation (100 percent) at night, so there is dew formation; however, values can drop to around 65 percent with the warmer daytime temperatures.

The microclimate of tropical rain forest stands is even more constant than the regional climate. Daily temperature variations of less than three degrees and relative humidity that remains above 90 percent make the microclimate within tropical rain forest stands one of the most constant of terrestrial environments. Of course, there are variations from the forest floor to the canopy; relative humidity decreases, and wind, temperature fluctuations, and solar radiation increase. High amounts of light penetrate the tallest tree layer, but denser, lower strata may prevent all but 0.05 percent from reaching the forest floor.

Rain forest soils belong to the Laterite Great Soils Group. They are so old that weathering is very deep and may exceed 20 meters. Laterization involves the leaching of silica (in the form of silicates), which leaves oxidized iron and aluminum. Laterite soils are very poor in nutrients; their clay particles have a low capacity to retain mineral ions, and soil organic matter is relatively low because of rapid decomposition. Minerals released during decomposition are quickly absorbed by root systems, so the ecosystem's nutrients are "bound up" in the biomass, especially in vegetation.

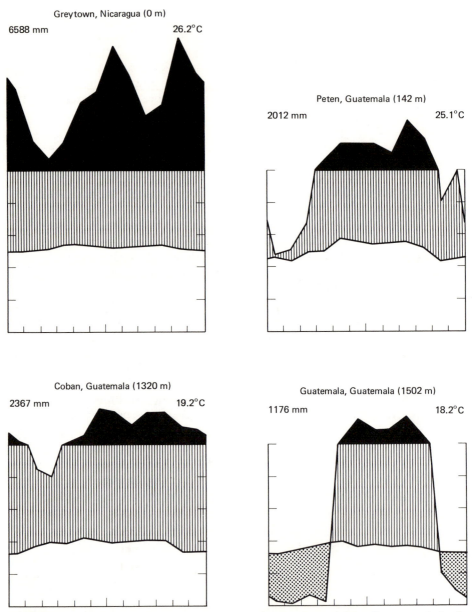

FIGURE 13.2 Climate diagrams in the tropical vegetation region: Greytown is in the Lowland Tropical Rain Forest region; Greytown, Peten, and Coban illustrate an elevational gradient; Guatemala is in the seasonal tropical forest region. *Source.* Redrawn from Walter, H. and H. Lieth. 1967. Klimadiagramm-weltatlas. Fischer, Jena, East Germany.

Additional environmental factors considered in other formations were fire, topographic relief, and the biotic factor. Fires are rare in undisturbed tropical rain forests, but are more common in areas of extensive human disturbance. Relief is relatively flat in the regions of this forest type. Some aspects of the biotic factor will be discussed later in this chapter.

Vegetation

Structure.   The vertical structure of a Lowland Tropical Rain Forest is so complex that it appears to be incomprehensible. However, a systematic plotting of plant heights reveals an arrangement as shown in Figure 13.3. The top tree stratum, the protruding layer, consists of sporadic individuals above the dense lower layers. The trees of the protruding layer are typically around 50 meters in height with spreading but shallow crowns. The second tree layer is more continuous than the first, but still has many gaps. Here the trees average 25 meters in height and have crowns that are nearly as deep as they are wide. The third tree layer is the densest of all the strata. These trees average 10 meters in height and have conical, tapering crowns. Life is concentrated in these upper layers.

The shrub, herb, and surface strata are very discontinuous. Their sparsity is probably a result of low light intensity beneath the three tree layers. The

FIGURE 13.3   Profile diagram of the tree strata of a Lowland Tropical Rain Forest. *Source*. Beard, J. S. 1946. The mora forests of Trinidad, British West Indies. Journal of Ecology 33:173–192. Published by Cambridge University Press.

openness of these strata gives a structure quite different from that of an impenetrable jungle; dense undergrowth is typical only of seral stands.

Composition.   Tropical rain forest stands are characterized by extremely high species diversity. A typical hectare of a rich deciduous forest in the eastern United States would contain about 12 tree species, having at least one individual 30 centimeters or more in diameter. A hectare in a Lowland Tropical Rain Forest would have at least 40 and, occasionally, up to 100 tree species. With such a large number of species, the adult individuals of any single species are very low in density and frequency. Thus, it is impossible to recognize dominants in most rain forests. Stands that do have dominant species typically occupy only small areas on special soils or habitats.

It has been hypothesized that the intense predation of seeds and seedlings reduces population densities and frequencies and, consequently, provides resource opportunities for other species. This would account for the high species diversity and the lack of dominants in tropical rain forests. Conversely, less efficient plant predation permits the dominance of a few species in temperate zone areas.

In spite of the great floristic diversity, growth-form similarity gives Lowland Tropical Rain Forests a rather monotonous appearance. The trees are virtually all evergreen angiosperms with slender, straight trunks branched only at the top. Their bases are frequently flared out as plank buttresses or, in the case of some subcanopy species, have stilt roots (Figure 13.4). Crown shape depends on the canopy level (page 231). Bark is generally smooth, thin, and light-colored. Leaves are usually large, ranging from 20 to nearly 200 square centimeters in area and from 8 to 24 centimeters in length. Except for the legumes (Leguminosae; commonest in the upper strata), the leaves are entire and frequently have a pointed apex known as a "drip tip." They are generally dark, dull green in color with a smooth surface and thick cuticle.

There is no synchronous leaf fall; individual species have a variety of periodicities. Some species drop all of their leaves at once. Others lose from different branches at different times. Any regular periodicities are most likely triggered by the small, annual variation in day lengths. The average life for individual leaves is about 13 to 14 months. The lack of climatic seasonality is also reflected in the absence of annual growth rings. However, tree age estimates based on annual increases in size indicate that most mature individuals are 200 to 250 years old. Large trees that die are replaced by different species (because of low species densities). This results in cyclic changes in undisturbed forests.

Lianas are a growth-form that is common in many tropical forest types. Lianas are vines that root in the soil and have flexible stems supported by trees. Their crowns are in the gaps of the tree strata (Figure 13.5). This

FIGURE 13.4 Buttress roots (top) and stilt roots (bottom) on tropical trees in Jamaica. The buttresses are approximately 3 m tall; the stilt roots are about 6 m tall.

growth-form is present in other forest formations (e.g., the deciduous forest has vines of wild grapes, *Vitis* spp., and poison ivy, *Rhus radicans*); however, lianas are commonest in the tropics. Ninety percent of liana species have a tropical distribution. They are especially prominent in the lower two tree strata and along the edge of forest stands. In some areas lianas are so dense that it is nearly impossible for trees to fall.

FIGURE 13.5   The leaves in the center of the photograph are from a liana; they fill a gap in the canopy.

Lianas have various means of climbing: scramblers rely on passive assistance by spines and thorns, twiners curl their stem tips about the tree parts, rootclimbers produce modified aerial rootlets that attach themselves to the trees, and tendril climbers have specialized shoot structures that curl around holdfasts, such as tree branches.

Epiphytes are another growth-form that is widespread in the tropics. As "air-plants," they are not rooted in the ground but are attached to the limbs and trunks of trees (Figure 13.6). Previously mentioned epiphytes include spanish moss *(Tillandsia usneoides)* of the Southern Mixed Hardwoods deciduous forest (Figure 9.13) and aerial lichens of the Boreal coniferous forest. True epiphytes are not parasitic but are photosynthetic and capable of growing on rain water and the nutrients dissolved in it.

Epiphytes may be classified as sun or shade plants, depending on their habitat preference. Most of the shade plants are ferns; they require high humidity to avoid desiccation and, thus, are commonest in the lowest tree layer. Most sun epiphytes are members of the orchid or bromeliad families (Orchidaceae and Bromeliaceae); they are tolerant of some desiccation and occur in the higher strata. Epiphytes show little uniformity in morphology. Adaptations will be discused later in this chapter.

A subdivision of the epiphytes is the epiphyll group—plants that grow on the leaf surfaces of other plants. Epiphylls include lichens, algae, liverworts, and bryophytes; even the germinated seeds of angiosperms can sometimes be found on leaf surfaces. The detailed study of epiphyll commu-

FIGURE 13.6 Epiphytes on a tree limb.

nities has revealed a successional pattern. Bacteria and fungi are also found on leaf surfaces.

Hemiepiphytes are intermediate between lianas and epiphytes. They include species that initially grow as lianas but lose their roots, and species that initially grow as epiphytes but form roots that reach the soil. The latter group includes the strangler fig *(Ficus aurea)*, whose roots clasp around the trunk of a supporting tree and prohibit increases in diameter. When the tree dies, it is, in effect, replaced by the strangler fig.

Typical shrub and herb growth-forms are relatively unimportant in tropical rain forests. Floristically, the forest is poor in herbaceous species; only a few angiosperm (flowering plant) families are present. Most of the herbs are ferns and fern-allies (especially small club mosses, *Selaginella* spp.). The growth of tropical herbs is of interest in that some species reach tall heights, even over 6 meters, and become part of the shrub or the lowest tree layer.

Production. The Lowland Tropical Rain Forest has greater annual net productivity than any other type of terrestrial vegetation in the world (Table 3.1). The total leaf area of a rain forest may be approximately the same or only somewhat more than that of a temperate deciduous forest. Also, the rates of carbon dioxide assimilation per unit area are the same in the two forests; therefore, the key factor is a 365-day growing season.

Since Laterite soils are very poor in nutrients, continuing high production is dependent on rapid, closed nutrient cycling. Lowland Tropical Rain

Forests produce more litter than any other type of vegetation. Litter decomposition is very rapid, as a result of continuously warm temperatures and high humidities. Plant roots absorb nearly all released nutrients. Little is leached from the soil. Evidence for this is that rivers draining tropical rain forest regions have low nutrient levels. Precipitation inputs compensate for the small nutrient losses. Rain water passing through the forest canopy also dissolves and carries nutrients that are released by leaves and epiphytes. The epiphytes can include nitrogen-fixing bacteria, which are growing as epiphylls, as well as nitrogen-fixing blue-green algae, contained in lichens. In some tropical rain forests it appears that the movement of nutrients by throughfall and stemflow is more important than movement by leaf fall.

The source of the original nutrient "capital" is unknown. The most commonly accepted hypothesis proposes that nutrients were accumulated in the biomass as the original soil-parent material was weathered during soil formation. This accumulation continued until the depth of the weathering exceeded the depth of the roots. Now the vegetation must rely upon the rapid, tight recycling system described in the previous paragraph. One of the great problems with the management of tropical rain forests is that disturbance destroys this system and results in the loss of the nutrient capital.

VEGETATION GRADIENTS IN MOUNTAINS

The many mountainous regions in tropical and subtropical North America have highly varied vegetation. Tropical mountain vegetation, like that of temperate regions, shows elevational zonation. From low to high elevation, tall tropical mountains have Lower Montane Rain Forest, Montane Rain Forest (also referred to as Cloud Forest), Montane Thicket or High Mountain Forest, Elfin Woodland, Páramo, and typical Alpine tundra vegetation (Figure 13.7).

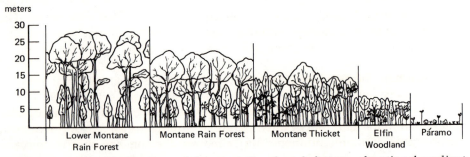

FIGURE 13.7   Profile diagrams of the vegetation found along an elevational gradient in tropical mountains. Alpine tundra vegetation of the type discussed in Chapter 7 may occur upslope from Páramo vegetation. *Source.* Beard, J. S. 1955. The classification of tropical American vegetation-types. Ecology 36:89–100.

Precipitation is variable in the mountains, but greatest values generally occur at around 2500 meters. This also is the elevation where cloud or fog cover usually is present, atmospheric humidity values are at a maximum, and solar radiation is least intense. Of course, temperatures decrease at higher elevations, but temperature variations are smallest in the cloud belt. Low temperatures become critical over 3000 meters. Soil differences parallel the elevational variations in climate. Tropical mountain soils include Podzols and other soil types, but not the Laterites usually associated with the tropics. Soil organic matter is greater at higher elevations where the rate of decomposition is slow.

The elevational changes in the environment bring about changes in the vegetation. In comparison to the Lowland Tropical Rain Forest, the forests become shorter with increasing elevation. Structure is simplified, in that the protruding layer is lost and the second tree layer decreases in height until it merges with the third. The vegetation also becomes more open at high elevations. The decrease in tree size is accompanied by decreases in leaf size and increases in root mats. Lianas initially increase in density, but they are not as common in the cloud belt and beyond. The number of epiphytes increases through the forested elevations, and lichens become relatively more important.

Lower Montane Rain Forest

The Lower Montane Rain Forest is found above elevations of 250 meters. The environment here is decidedly tropical and results in what is considered to be the maximum development of tropical rain forest vegetation. Atmospheric humidity values are constantly high because of stationary clouds around the mountains. Temperatures are still mild at this low elevation. Precipitation is very high, but the steep relief increases runoff. The soils are therefore better drained than Lowland Tropical Rain Forest soils, which are frequently waterlogged following rainfall. There are two tree strata: the canopy at 25 to 30 meters and the subcanopy at about 10 meters.

Montane Rain Forest (Cloud Forest)

The next higher elevational zone supports Montane Rain Forest (Cloud Forest) vegetation. It is perhaps the most interesting of the forest zones. Here, humidity, precipitation, fog, and cloud cover are at a maximum. Temperatures are generally cooler than in the Lower Montane Rain Forest by 12 to 20°C. This results in somewhat of a temperate climate; therefore, the name "Temperate Rain Forest" also is applied to this forest, especially toward the northern part of its range.

In general, most of the Montane Rain Forest resembles a tropical forest; however, there are distinct differences, especially floristically. Tree strata occur at around 10 and 20 meters. There are fewer lianas and more epiphytes than at lower elevations. The primary plants of the shrub layer are tree ferns and dwarf palms. Mosses are common; their retention of moisture contributes to the constant high humidity within the forest. In Mexico, the northern portion of this forest is dominated by pines (*Pinus* spp.) or oaks (*Quercus* spp.). Many temperate hardwoods are present, including genera and species found in the eastern United States; however, an understory of tropical species relates this forest to the tropics.

Montane Thicket or High Mountain Forest

The next higher elevational belt is that of the Montane Thicket or High Mountain Forest. The Montane Thicket is a low forest with trees of 10 to 15 meters and very little understory. It is found on lower mountains.

The High Mountain Forest occurs on taller peaks; therefore, it is well developed in Mexico and Central America. It occupies a cool belt above the frost line and thus is dominated by genera that are common further north. The chief dominants in much of this zone are pines (which are increased by fire) and oaks, along with species of such temperate genera as alder (*Alnus* spp.) and maple (*Acer* spp.). Thus, there is a strong relationship with the flora of the eastern United States. The High Mountain Forest community at timberline in Mexico (an elevation around 4200 meters) is dominated by pines and firs (*Abies* spp.). Elsewhere, the zone of High Mountain Forest supports vegetation resembling that of the southern Rocky Mountains, in that Pinyon-Juniper and scrub oak stands are present. Despite these affinities with temperate vegetation, the understory of the High Mountain Forest is of a tropical nature.

Elfin Woodland

The Elfin Woodland is the highest tree-dominated zone in truly tropical mountains. It is found at an elevation where temperatures average between 15 and 20°C. Cloud cover reduces solar radiation by about 40 percent, and humidity values are very high. The forest has a single tree stratum at 1 to 10 meters that combines with an understory of dwarf palms and tree ferns to make stands nearly impenetrable. The trees have a gnarled, twisted shape with long branches extending away from the wind.

It is thought that the shape of the trees is the result of nutrient deficiencies produced as slow transpiration rates reduce the root absorption of water and minerals. Transpiration is slowed by high humidity, fog, and low temperatures. One adaptation that is thought to result in increased transpiration is the rounded dense canopy of the trees. This shape reduces cooling wind flow through the crowns, and the warm temperatures maintain higher transpiration rates. Another adaptation of many trees is stilt roots or dense mats of roots on the soil surface, where oxygen is more available than in the continually wet soil. Lianas are less common than they are downslope. This also may be related to the problems associated with maintaining a flow of water and nutrients from root systems to the canopy, despite slow transpiration. Mosses and lichens are common as epiphytes; such abundance is a sign of high atmospheric humidity.

Trees are absent above the Elfin Woodland. The causes of tropical timberlines are unclear; however, in comparison to the forested zones, the environment at this elevation (and higher) is harsh. For example, precipitation is greatly decreased, and a daily freezing-thawing cycle can occur throughout most of the year. Precipitation may be the key factor that determines timberline. Evidence for this is the fact that trees occur at highest elevations along streams, even though these are sites of cold air drainage.

Páramo

The Páramo, a type of tropical alpine vegetation, occurs in some Caribbean islands and as far north as Costa Rica. Its elevational zone is virtually a cold desert; temperatures and precipitation are low. However, the Páramo environment is different from that of the Alpine tundra discussed in Chapter 7; its location at low latitudes results in a longer growing season, milder winters, and less intense winds. Little ecological work has been done in the Páramos of Middle America. In South America, Páramos have a low herb layer dominated by grasses (Gramineae) and forbs. The grasses are frequently sclerophyllous bunch grasses of one-half to one meter in height and width. A second, taller layer of columnar rosette plants is frequently present. These plants reach two or more meters and may have a treelike growth-form. Most of them are members of the sunflower family (Compositae). The presence of such tall plants is evidence that the environment is milder than in other alpine areas.

Alpine Tundra

Alpine tundra vegetation, similar to that described in Chapter 7, occurs in areas with a more severe environment—areas above the Páramo zone and in subtropical mountains where Páramo vegetation is not found.

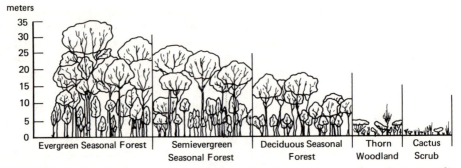

FIGURE 13.8 Profile diagrams of the vegetation found along a climatic seasonality gradient. Desert vegetation may be found under more arid conditions than occur in Cactus Scrub vegetation. *Source*. Beard, J. S. 1955. The classification of tropical American vegetation-types. Ecology 36:89–100.

VEGETATION GRADIENTS WITH INCREASING CLIMATIC SEASONALITY

Some tropical regions have a seasonal climate with distinct wet and dry periods. The severity of the dry seasons varies from location to location. Along a gradient of increasing severity are Evergreen Seasonal Forests, Semievergreen Seasonal Forests, Deciduous Seasonal Forests, Thorn Woodlands, Cactus Scrub, and desert vegetation (Figure 13.8). This section is concerned with the first four of these types of vegetation; their primary environmental gradient is the increasing duration of the dry period. The Cactus Scrub and desert belong to the desert formation; they are differentiated by the duration of drought and also by total annual precipitation.

Some general trends in vegetation from the Evergreen Seasonal Forest to the desert include decreases in the following: number and height of tree strata, height of tree branching, tree diameter, frequency of buttressing, leaf size, ground vegetation, density of epiphytes, and total species diversity. There are increases in the deciduousness of upper strata (except for the Thorn Woodland) and the presence of spines. Lianas initially increase but then decrease along this gradient.

Evergreen Seasonal Forest

Tropical rain forest vegetation can be found in regions with drought that lasts a few weeks, provided that total precipitation is very high. In areas with a longer dry period, there is greater seasonal activity of the vegetation; the forest that occurs with somewhat longer drought is the Evergreen Seasonal Forest.

This forest type annually receives more than 180 centimeters of precipitation and has a three-month dry season with only 5 to 10 centimeters. There are three tree strata. The protruding layer is very open, the mid stratum is more or less continuous, and the lowest layer is open. All three tree layers are lower in height than in a Lowland Tropical Rain Forest. About 25 percent of the species of the protruding layer are drought deciduous; these species may be abundant. Trees of the lower strata are less exposed to drought and are characteristically evergreen. Large trees are scattered, so the forest differs in appearance from a massive rain forest. The trees have low branches; lianas and epiphytes are common; and the ground vegetation is denser than in a rain forest.

Semievergreen Seasonal Forest

The Semievergreen Seasonal Forest is characteristic of regions with a total precipitation of 80 to 150 centimeters and a five-month drought period with 2.5 to 10 centimeters. This vegetation has lower species diversity than the Evergreen Seasonal Forest, and its stands are more often dominated by one or a few species. There are two major tree strata: an upper closed canopy layer of deciduous species and a lower layer consisting mostly of evergreens. The deciduous species lose their leaves following the onset of drought. About one month before the rainy season, new leaves begin to form, apparently triggered by increasing temperatures. Some species are facultative evergreens, in that they are able to retain their leaves in years with higher than normal precipitation. The trees of Semievergreen Seasonal Forests are smaller than those of the Evergreen Seasonal Forest. They have low branches. A few woody species have thorns. Epiphytes, especially ferns and mosses, are rare, but lianas are more common than in the other seasonal forests.

Deciduous Seasonal Forest

The Deciduous Seasonal Forest has five months with precipitation under 10 centimeters and two months with precipitation under 2.5 centimeters. This forest is even poorer in species diversity; one species may dominate individual stands. The two tree strata are an open upper layer of mostly deciduous trees (here they are obligately deciduous) and a dense lower layer of evergreens. The trees are small and have low branches; some have spines. Lianas and epiphytes are uncommon. Ground cover is poorly developed.

Thorn Woodland

The Thorn woodland is found in regions with a seasonal, semiarid climate. The vegetation varies from being somewhat open to very dense. The single tree layer is scrubby and has low species diversity. Most of the trees have spines and small evergreen xeromorphic leaves.

Dry Evergreen Forest

A type of vegetation related to this gradient of seasonal forests is the Dry Evergreen type. It is dominant in regions where precipitation is low throughout the year. It is especially common on soils with rapid drainage; many stands are located in coastal regions with porous bedrock. Dry Evergreen vegetation has a canopy, protruding individuals, and an open understory. The trees are typically sclerophyllous and are smaller in size where moisture availability is low.

TROPICAL SAVANNA VEGETATION

The term savanna is frequently restricted to tropical vegetation that consists of a dense layer of xeromorphic grasses and sedges (*Carex* spp.), with or even without a very open layer of short trees or shrubs (Figure 13.9). Such

FIGURE 13.9   Tropical Savanna vegetation in southcentral Jamaica.

savannas are scattered throughout much of Central America. They are best developed on the west side of the continent, but nowhere do stands reach the extent of those in South America or Africa.

Several hypotheses have been proposed to explain the origin of tropical savannas: climate, human influence, and soils. There is evidence to discount both climate and human influence as the key factor. For example, although most savannas of Central America occur in strongly seasonal climates, they have a patchy distribution and, in addition, are present in a variety of climatic regions. Also, although human clearing and burning of forest vegetation has produced some savannas, the majority are strongly related to particular habitat factors. Since one of those habitat factors is soil, that appears to be the best explanation for the origin and existence of most savannas.

Tropical savannas can be classified as nonseasonal, seasonal, or hyperseasonal. Regions with continually wet climates have nonseasonal savannas. These occur on the driest and the most nutrient poor soils, that is, those that are well drained, deep, medium to coarse in texture, and have a water table well below the root zone. These stands are thought to be relicts from a drier climatic period when savanna vegetation dominated the landscape. Increased precipitation brought about the development of rain forest vegetation on all but the poorest soils.

Seasonal savannas also occur on dry, poor soils, but in areas of a regional climate with distinct wet and dry periods. Human influence has been important in these areas; the increased frequency of fires maintains the savannas by stopping the spread of forests. Savanna vegetation is well adapted to fires; the trees have thick, fire-resistant bark and most of the herbs are perennials.

Hyperseasonal savannas are the most common type in Central America. They occur in areas of flat relief, where the effects of seasonality are pronounced by heavy soils that have a relatively impermeable subsurface layer. During periods of high precipitation, this layer produces a perched water table; this results in an overabundance of soil water. During dry seasons, the heavy soil texture results in limited soil water availability.

ADAPTATIONS

A variety of growth-form adaptations have already been mentioned. This section elaborates on some of these and introduces others.

Many tropical trees have a buttress base, especially in areas of high precipitation (Figure 13.4). Such a growth-form has long been conjectured to aid in support, a factor of importance since most tropical trees have very shallow root systems, another adaptation. Roots of tropical trees are primarily in the upper meter of soil. The greatest amount of fine absorbing roots are at or only slightly below the litter layer, a position that is advantageous for absorption of mineral nutrients released through decomposition. Furthermore, mycorrhizae form a direct connection between tree roots and decomposing organic matter.

The growth-form of tropical leaves has also been a subject of conjecture. Drip tips and smooth surfaces appear to facilitate the rapid shedding of water (Figure 13.10); this could reduce the leaching of nutrients from leaves. It also may promote transpiration and thereby increase the rate of absorption of mineral nutrients. A thick leaf cuticle is the major reason for the smooth leaf surface. It also is an adaptation to the dry conditions that occur in the tree canopy for a few hours every day as a result of bright sunlight and high temperatures. Other adaptations of leaves to the canopy environment include small size, short drip tip, vertical orientation, and midday closure of stomates. Trees of high elevation montane forests, where transpiration and nutrient uptake rates are slow, commonly have larger leaves with more numerous stomates.

FIGURE 13.10   Drip tips on the leaves of tropical species.

The reproductive patterns of tropical trees are also highly adaptive. Cauliflory—the development of flowers on old woody trunks and limbs—is common in deciduous understory species. It is thought that this increases the visibility of flowers to pollinators, an important factor in dense forests. Also, many tree species do not form seeds every year but produce them in large amounts more infrequently. The large amounts alone increase the probability that some seeds will escape predation by primary consumers. Other species have low seed predation, because their seeds are poisonous or bitter tasting. Seeds of tropical species also usually contain large amounts of stored food, a factor which aids initial growth in the low light intensity on the forest floor. Finally, as in all forests, but especially in the structurally complex tropical forests, tree seedlings must be able to phenotypically adjust to the different microenvironments that they encounter as they grow from seedlings to tall, mature individuals.

Tropical lianas have fairly high light requirements. Thus, they are most common in open forests, clearings, or along edges of dense stands. It is thought that they are more common in the tropics than in any other region, because water transport is not a great problem in wet, warm environments. They have very large vessel elements (water-conducting cells) in their stems.

Epiphyte adaptations include light, wind-dispersed seeds (or spores) and bird-dispersal mechanisms. Many epiphytes accumulate organic matter around their root holdfasts; sometimes this is aided by ant colonies that are associated with the roots. The organic matter retains moisture and is a source of mineral nutrients. The ants may provide significant amounts of nitrogen. Most bromeliad epiphytes have rosettes of overlapping cup-shaped leaves that are usually lacking in terrestrial members of the same family (Figure 13.11). These leaves catch and retain organic matter and water with dissolved minerals. The bases of the leaves have absorbing scales. It has been proposed that epiphytes are more common on old trees than young ones, because older individuals release more minerals in leaf throughfall and stem flow than do young rapidly growing trees.

Of all tropical forests, epiphytes are most common in the montane forests. It appears that frequent rains are more important in their distribution than is total annual precipitation. Many epiphytes are able to store water; some are succulents and others have thick rhizomes. Low rates of transpiration are common in epiphytes. Some species shed leaves during dry periods, and a few orchids lack stems and leaves (their root system is photosynthetic). Other epiphytes have a thick cuticle and root structures that retard transpiration. Some species are capable of dark fixation of carbon dioxide, and therefore their stomates are closed during the day. Still others can survive up to 50 percent dehydration.

Lastly, epiphytes, especially shade epiphytes, are adapted to low light intensities. This is also true for herbs on the forest floor.

FIGURE 13.11   The bromeliad epiphyte *Gravisia aquilega* in Trinidad. (*a*) and (*b*) Its habitat. (*c*) A single individual with humus and water splashed from the cups of the overlapping leaves. (*d*) View of an individual with the leaf blades cut away to show the cup areas between the leaf bases. *Source*. Pittendrigh, C. S. 1948. The bromeliad-*Anopheles*-malaria complex in Trinidad. I—The bromeliad flora. Evolution 2:58–89.

HUMAN IMPACT

Much tropical vegetation—forests in particular—has been cleared for agriculture. The system of agriculture that developed in the tropics is different from that of temperate regions. It is referred to as "slash and burn" or "swidden" agriculture. An area within a forest is cut, and the downed vegetation is burned. This burning and the decomposition of organic matter adds nutrients to the soil. Crops are planted and harvested. Within a few years, leaching makes the soil too poor for productive agriculture, and the cleared area is rapidly invaded by weedy secondary species that are very difficult to remove. At that point, the site is abandoned and another area in the forest is cut, burned, and tilled until it, too, becomes unproductive. In the past, abandoned land was left fallow for 30 to 40 years or more before it was recut and reused.

It has been said that shifting agriculture can support about 8 people per square kilometer. Thus, the impact of the aboriginal Indians, whose populations were small, was minimal (except in the Yucatan Peninsula where Mayan corn agriculture reduced the original forests to secondary scrub forests). However, in more recent times foreign demand for tropical agricultural products and large population increases in Central America led to changes in this relatively stable agricultural system. Larger areas were cleared, and abandoned land was left fallow for shorter periods of time before being reused. This led to soil deterioration and progressively poorer forests. Agriculture also spread to increasingly higher elevations where erosion became severe unless extensive stone terracing was built.

With only a low percentage of the land arable by modern temperate zone techniques, in the future it appears that shifting agriculture must be replaced by small-scale (family size) agriculture that will involve the growth of several crops on a single field. To a degree, this type of agriculture would then resemble the structure and mineral cycling of the original vegetation.

Although the agricultural demands of growing populations have had the greatest destructive effect on tropical forests, timber utilization has also been influential. Most of the wood that is harvested is used as fuel for cooking, heating, and local industries. However, the cutting of timber for lumber production has also had a great impact on the tropical landscape, especially in previously undisturbed areas.

Until World War II, most of the logging consisted of selective cutting of valuable trees such as mahogany (*Swietenia* spp.), so the forest ecosystems were left relatively intact, except where agricultural expansion had followed. Lumber shortages and the high prices of wood imported from the United States and Canada led to the greater use of local forests, including species formerly not cut. Clear-cutting—the logging of entire stands—resulted in the loss of soil nutrients and problems with erosion. Exploitation of remote areas was assisted greatly by the construction of the Inter-American Highway through Central America.

In spite of the awareness of the problems created by the logging of tropical forests, most Central American countries cannot financially afford to limit the exploitation. Also, the lack of scientific knowledge in reforestation of tropical regions has hindered the development of a sustaining forest industry, though relatively recently there have been improvements. Introduced tree species have been tested, and there is a trend toward reforestation with fast-growing temperate conifers.

In areas left to natural succession, it is feared that forests of the original composition and structure will not return. Lack of nutrients, loss of mycorrhizal fungi, and continuing human disturbance may ensure this. Also, because of the low densities and frequencies of most individual species, disturbance frequently results in local extinctions. Not only are some species lost, but the ecological interactions important to others are lost as well; the long-range results of these changes are impossible to predict. At the present rate of cutting, there will be little undisturbed tropical rain forest left anywhere in the world by the end of this century. Since this vegetation is the site of great genetic and ecological diversity, its loss would be a tragic event, unparalleled in history.

SUGGESTED READINGS FOR FURTHER STUDY

Asprey, G. F. and R. G. Robbins. 1953. The vegetation of Jamaica. Ecological Monographs 23:359–412.

Beard, J. S. 1944. Climax vegetation in tropical America. Ecology 25:127–158.

———. 1953. The savanna vegetation of northern tropical America. Ecological Monographs 23:149–215.

———. 1955. The classification of tropical American vegetation-types. Ecology 36:89–100.

Bourliere, F. and M. Hadley. 1970. The ecology of tropical savannas. Annual Review of Ecology and Systematics 1:125–152.

Craighead, F. C., Sr. 1971. The trees of south Florida, Vol. 1. The natural environments and their succession. University of Miami Press, Coral Gables, Florida. 212 p.

Flores, M. G., L. J. Jimenez, S. Z. Madrigal, R. F. Moncayo, and T. F. Takaki. 1971. Tipos de vegetación de la República Mexicana. Subsecretaria de Planeacion Direccion General de Estudios, Director Agrologia, SRH, Mexico, D. F. 59 p. + map.

Golley, F. B. and E. Medina (eds.). 1975. Tropical ecological systems: Trends in terrestrial and aquatic research. Springer-Verlag, New York. 398 p.

Grubb, P. J. and E. V. J. Tanner. 1976. The montane forests and soils of Jamaica: A reassessment. Journal of the Arnold Arboretum 57:313–368.

Holdridge, L. R. 1956. Middle America, p. 183–200. *In* S. Haden-Guest, J. K. Wright, and E. M. Teclaff (eds.). World geography of forest resources. Ronald, New York.

Holdridge, L. R., W. C. Grenke, W. H. Hatheway, T. Liang, and J. A. Tosi, Jr. 1971. Forest environments in tropical life zones: A pilot study. Pergamon, Oxford, 747 p.

Janzen, D. H. 1971. Seed predation by animals. Annual Review of Ecology and Systematics 2:465–492.

———. 1975. Ecology of plants in the tropics. Edward Arnold, London. 66 p.

Jordan, C. F. and J. R. Kline. 1972. Mineral cycling: Some basic concepts and their application in a tropical rain forest. Annual Review of Ecology and Systematics 3:33–50.

Leigh, E. G., Jr. 1975. Structure and climate in tropical rain forest. Annual Review of Ecology and Systematics 6:67–86.

Leopold, A. S. 1950. Vegetation zones of Mexico. Ecology 31:507–518.

Longman, K. A. and I. J. Jenik. 1974. Tropical forest and its environment. Longman, London. 196 p.

Odum, H. T. and R. F. Pigeon (eds.). 1970. A tropical rain forest: A study of irradiation and ecology at Elverde, Puerto Rico. United States Atomic Energy Commission, Oak Ridge, Tennessee.

Pittendrigh, C. S. 1948. The bromeliad-*Anopheles*-malaria complex in Trinidad. I— The bromeliad flora. Evolution 2:58–89.

Richards, P. W. 1952. The tropical rain forest. Cambridge University Press, London. 450 p.

———. 1973. The tropical rain forest. Scientific American 229(6):58–67.

Verdoorn, F. (ed.). 1945. Plants and plant science in Latin America. Chronica Botanica Co., Waltham, Massachusetts. 381 p.

Vogt, W. 1948. Latin America timber ltd. Unasylva 2:19–25.

Walter, H. 1971. Ecology of tropical and subtropical vegetation. Translated by D. Mueller-Dombois. Edited by J. H. Burnett. Oliver & Boyd, Edinburgh. 539 p.

APPENDIX

SPECIES NAMES FROM PART 2

acacias	*Acacia* spp.
agaves	*Agave* spp.
alders	*Alnus* spp.
alpine azalea	*Loiseleuria procumbens*
American basswood	*Tilia americana*
American beech	*Fagus grandifolia*
American chestnut	*Castanea dentata*
American elm	*Ulmus americana*
arctic meadow-rue	*Thalictrum alpinum*
ashes	*Fraxinus* spp.
azaleas	*Rhododendron* spp.
bald cypress	*Taxodium distichum*
balsam fir	*Abies balsamea*
basswoods	*Tilia* spp.
bay	*Umbellularia californica*
bearberry willow	*Salix uva-ursi*
beeches	*Fagus* spp.
big bluestem	*Andropogon gerardi*
birches	*Betula* spp.
bitternut hickory	*Carya cordiformis*
blackjack oak	*Quercus marilandica*
black oak	*Quercus velutina*
black spruce	*Picea mariana*
blueberries	*Vaccinium* spp.
bluebunch wheat grass	*Agropyron spicatum*

blue grama	*Bouteloua gracilis*
blue oak	*Quercus douglasii*
bristlecone pine	*Pinus aristata*
brittlebush	*Encelia farinosa*
broad-leaved lungwort	*Mertensia ciliata*
brome grasses	*Bromus* spp.
buffalo grass	*Buchloë dactyloides*
bur oak	*Quercus macrocarpa*
bur sage	*Franseria dumosa*
buttercup (tundra)	*Ranunculus glacialis*
California buckeye	*Aesculus californica*
California white oak	*Quercus lobata*
Canada wild-rye	*Elymus canadensis*
canyon live oak	*Quercus chrysolepis*
ceanothus	*Ceanothus* spp.
chamise	*Adenostoma fasciculatum*
chestnut	*Castanea dentata*
chestnut oak	*Quercus prinus*
club mosses	*Selaginella* spp.
coast live oak	*Quercus agrifolia*
cottongrass	*Eriophorum* spp., including *E. vaginatum*
cottonwoods	*Populus* spp.
coulter pine	*Pinus coulteri*
creosote bush	*Larrea divaricata*
diapensia	*Diapensia lapponica*
digger pine	*Pinus sabiniana*
douglas fir	*Pseudotsuga menziesii*
downy chess	*Bromus tectorum*
downy oat-grass	*Trisetum spicatum*
dwarf birch	*Betula nana* ssp. *exilis*
eastern hemlock	*Tsuga canadensis*
elms	*Ulmus* spp.
Engelmann spruce	*Picea engelmanii*
evergreen magnolia	*Magnolia grandiflora*
fescues	*Festuca* spp.
firs	*Abies* spp.
flowering dogwood	*Cornus florida*
Fraser fir	*Abies fraseri*
Gambel's oak	*Quercus gambelii*
giant sequoia	*Sequoiadendron giganteum*
goldenrods	*Solidago* spp.
grama grasses	*Bouteloua* spp.
greasewood	*Sarcobatus vermiculatus*

hairy grama	*Bouteloua hirsuta*
hickories	*Carya* spp.
Idaho fescue	*Festuca idahoensis*
incense cedar	*Calocedrus decurrens*
Indian grass	*Sorghastrum nutans*
interior live oak	*Quercus wislizenii*
jack pine	*Pinus banksiana*
Jeffrey pine	*Pinus jeffreyi*
Joshua tree	*Yucca brevifolia*
Junegrass	*Koeleria cristata*
junipers	*Juniperus* spp., including *J. monosperma*, *J. occidentalis*, *J. osteosperma*, and *J. scopulorum*
larch (tamarack)	*Larix laricina*
laurel oak	*Quercus laurifolia*
little bluestem	*Andropogon scoparius*
live oak	*Quercus virginiana*
loblolly pine	*Pinus taeda*
lodgepole pine	*Pinus contorta*
longleaf pine	*Pinus palustris*
madrone	*Arbutus menziesii*
mahoganies	*Swietenia* spp.
manzanitas	*Arctostaphylos* spp.
maples	*Acer* spp.
mesquites	*Prosopis* spp., including *P. juliflora*
milkweeds	*Aesclepias* spp.
mockernut hickory	*Carya tomentosa*
Mojave yucca	*Yucca schidigera*
mountain avens	*Dryas* spp.
mountain hemlock	*Tsuga mertensiana*
mountain mahogany	*Cercocarpus* spp.
mountain sorrel	*Oxyria digyna*
mouse barley	*Hordeum murinum*
needle-and-thread grass	*Stipa comata*
needlegrass	*Stipa spartea*
Nuttal saltbush	*Atriplex nuttallii*
oaks	*Quercus* spp.
ocotillo	*Fouqueria splendens*
oneseed juniper	*Juniperus monosperma*
ox-eye	*Heliopsis helianthoides*
palo verdes	*Cercidium* spp.
paper birch	*Betula papyrifera*
peat mosses	*Sphagnum* spp.
pines	*Pinus* spp.

pinyon pines	*Pinus cembroides, P. edulis,* and *P. monophylla*
pitch pine	*Pinus rigida*
poison ivy	*Rhus radicans*
ponderosa pine	*Pinus ponderosa*
post oak	*Quercus stellata*
prairie cordgrass	*Spartina pectinata*
prairie dock	*Silphium terebinthinaceum*
prairie dropseed	*Sporobolus heterolepis*
prickly pear cacti	*Opuntia* spp.
purple needlegrass	*Stipa pulchra*
quaking aspen	*Populus tremuloides*
redbud	*Cercis canadensis*
red fir	*Abies magnifica*
red hickory	*Carya ovalis*
red maple	*Acer rubrum*
red oak	*Quercus rubra*
red pine	*Pinus resinosa*
red spruce	*Picea rubens*
redwood	*Sequoia sempervirens*
reindeer lichens	*Cladonia* spp.
resurrection plant	*Selaginella lepidophylla*
rhododendrons	*Rhododendron* spp.
sagebrush	*Artemisia tridentata*
saguaro	*Cereus giganteus*
sandberg bluegrass	*Poa secunda*
scrub oak	*Quercus turbinella*
sedges	*Carex* spp., including *C. aquatilis*
shadscale	*Atriplex confertifolia*
shagbark hickory	*Carya ovata*
shortleaf pine	*Pinus echinata*
side-oats grama	*Bouteloua curtipendula*
Sitka spruce	*Picea sitchensis*
slash pine	*Pinus elliottii*
Spanish moss	*Tillandsia usneoides*
spruces	*Picea* spp.
spurges	*Euphorbia* spp.
strangler fig	*Ficus aurea*
subalpine fir	*Abies lasiocarpa*
sugar maple	*Acer saccharum*
sugar pine	*Pinus lambertiana*
sweet buckeye	*Aesculus octandra*
switchgrass	*Panicum virgatum*
tamarack (larch)	*Larix laricina*

tarbush	*Flourensia cernua*
tobosa grasses	*Hilaria* spp.
tuliptree	*Liriodendron tulipifera*
tupelos	*Nyssa* spp.
Virginia pine	*Pinus virginiana*
western arbor vitae	*Thuja plicata*
western hemlock	*Tsuga heterophylla*
western wheat grass	*Agropyron smithii*
whitebark pine	*Pinus albicaulis*
white basswood	*Tilia heterophylla*
white fir	*Abies concolor*
white oak	*Quercus alba*
white pine	*Pinus strobus*
white spruce	*Picea glauca*
wild grapes	*Vitis* spp.
wild oats	*Avena fatua*
willows	*Salix* spp., including *S. pulchra*
winter fat	*Eurotia lanata*
wire grasses	*Aristida* spp., including *A. longiseta*
yellow birch	*Betula lutea*
yuccas	*Yucca* spp., including *Y. elata*

Index